TO   Anna-Karien

HAPPY GROWING !

designing
regenerative
food systems

Marina O'Connell

A powerful and personal book about transforming the land for the better. Marina O'Connell weaves inspirational stories of redesign and transformation, showing how regenerative methods for agriculture and food have come to life. In five years, she created a productive, diverse, profitable and regenerative farm from depleted soil, and has said, over here is a path, now we can walk it.

Professor Jules Pretty, author of *Regenerating Agriculture* (1995), *Agri-Culture* (2003) and *The East Country* (2017)

It is hard to overestimate how profoundly important the urgent and ambitious reimagining of food and farming systems is, how we might do so in ways that are regenerative, restorative and transformative. How to create farms and gardens that build soil, community, and possibility? This book is your opportunity to learn from a master. What Marina O'Connell has created at the Apricot Centre is nothing short of miraculous. In these precious pages she shares everything you need to know in order to do the same. May this book spark a revolution of the agricultural and horticultural imagination.

Rob Hopkins, co-founder of the Transition Towns movement and author of *From What Is to What If*

Marina O'Connell is a practical possibilist… a weaver of integrated agro-ecological systems that regenerate the soil under our feet and the food on our plates. She convincingly makes the case – in this most timely of books – for how agricultural practices that grew up on the fringes of our dominant industrial food system can help shift contemporary food culture to an agro-ecological paradigm that benefits both people and planet. Highly recommended for farmers and students needing an evidence-based toolkit for developing regenerative food systems.

Jonathan M. Code, Lecturer in Sustainable Land Management, Royal Agricultural University

Few people have attempted to survey the different strands of sustainable agriculture, but Marina has done so in a really valuable way, showing the links and crossovers between the different techniques along with some good science and other references.

Martin Crawford, Director, Agroforestry Research Trust

Hope for the future of humanity and wild nature lies not with governments, corporates and international conferences but with grassroots movements – especially in food and farming. Marina O'Connell is a farmer and an educator. This excellent regenerative farming design toolkit of what's already in train worldwide is just what's needed for enabling the coming agro-ecological revolution.

Colin Tudge, Co-founder of the Oxford Real Farming Conference and the College for Real Farming and Food Culture

The strands Marina has succeeded in weaving together – including the psychological support for people connected to the farm, the creative local marketing, and the collaboration with local businesses – all are wonderful threads in what I see as the establishment of a model local food economy.

Helena Norberg-Hodge, Founder and Director,
Local Futures

Nature works not just because it is diverse, but because it is made up of relationships between the many elements. So it is with regenerative agriculture. No single approach has all the answers, but by weaving them together, forming relationships, we create a whole systems approach that works for people and planet. Marina has done a great service for students, farmers and growers, and everyone with a passion to bring land and community to life. She has shown how we can connect disciplines and approaches, and bases this not just on theory, but on her extraordinary farm at Huxhams Cross. Highly recommended!

Andy Goldring, Permaculture Association

With this publication Marina O'Connell has done all food citizens a great service – of presenting the main sustainable farming approaches that have crystallised over the last century as a response to the industrial farming project, in a straightforward and accessible format without discrimination or ranking. The author – a seasoned farmer herself – provides an overview of each approach in turn: Biodynamic, Organic, Permaculture, Agroforestry, Agroecological and Regenerative Agriculture, in the context of the contemporary challenges of climate change, biodiversity loss, food security and human health, so that farmers and growers, students, policy makers, researchers and food citizens in general can grasp not only the main concepts but also ways to mitigate the challenges going forward. The author concludes that we might 'ferment' transformational change through being discerning in our own food choices, through healing our own traumas which our landscapes only reflect, through enabling access to land and through appropriate training provision. This book should be seen as a primer for everyone entering the food and farming debate as well as those who want to broaden their perspectives.

Julia Wright (PhD), Associate Professor, Centre for Agroecology,
Water and Resilience (CAWR), Coventry University, UK, and
Council Member of the Biodynamic Association UK

Marina O'Connell has helped our family farm become more sustainable. Within the partnership, she has helped redesign a 10-acre field into Biodynamic Conversion with a rare and valuable combination of deep experiential and practical insight, drawing on this regenerative toolkit for horticulture, farming and community life with the land. We value her skill in weaving together different approaches to Agroecology, including Permaculture, Organic practices and Biodynamic methods, which are so well explained in this timely handbook.

Dr Miche Fabre Lewin and Dr Flora Gathorne-Hardy,
Great Glemham Farms, Suffolk

Hawthorn Press

Published 2022 by Hawthorn Press, Hawthorn House
1 Lansdown Lane, Stroud, Gloucestershire, United Kingdom, GL5 1BJ
Tel: +44 1453 757040 Email: info@hawthornpress.com
www.hawthornpress.com

Cover design by Jason Conway
Cover photography by Jimmy Edmonds
Typesetting by Winslade Graphics
Printed by Cambrian Printers Ltd, Wales

Carbon-balanced at source and printed on uncoated FSC© certified paper using sustainable printing procedures. The cover makes use of a sustainable, biodegradable and recyclable lamination.

WORLD LAND TRUST™
www.carbonbalancedpaper.com
CBP006075

FSC
www.fsc.org
MIX
Paper from responsible sources
FSC® C004116

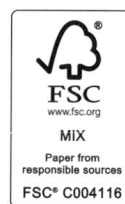

British Library Cataloguing in Publication Data applied for

ISBN 978-1-912480-54-8

designing
regenerative
food systems

and why we
need them
now

Marina O'Connell

Hawthorn Press

Glossary of terms

Many of the terms used in this book are interchangeable with other ones. I have defined my usage of these terms below. Not everyone will agree with my definitions, but for the sake of clarity, and until the terms settle into common usage, this is how they are used in this book.

Industrial farming is the form of farming generally thought of as 'conventional farming', in references to the fact that it is mainstream in the global north. However, it is a relatively new form of farming; only three generations of farmers have used these industrialised methods. Sociologists call this the 'industrial production paradigm'

Sustainable food systems is a collective name for all of the food systems described in this book, sometimes called 'biological', 'ecological' or 'alternative' farming systems. Sometimes 'agroecology' or 'regenerative' is used as an umbrella term for all of them. I have not used these two terms in this way, since each is presented in this book as a food system in its own right. The sociologists call this the 'ecologically integrated paradigm'.

Regenerative agriculture implies something more than sustainable agriculture – a system that repairs and rehabilitates the badly damaged soil and water systems.

Pesticide is used as a collective term for insecticides, fungicides and herbicides.

Farmers are people who produce food, whether on a large or a small scale. They're sometimes described as 'growers'. Many farmers are women, people of colour and/or indigenous people.

The agricultural revolution, which happened first in England, was the move from subsistence farming on common land, to enclosed, privatised food production that was carried out as a business for money. This started in the 14th century in England and is still happening today in many parts of the world.

The industrial farming revolution started in the global north in the early 20th century. It introduced nitrate fertilisers, pesticides, tractors, new varieties of crops, artificial insemination and battery farming. The focus was on yields.

The green revolution refers to the rolling out of industrial processes in farming across the global south from the 1960s onwards.

The sustainable farming revolution refers to the changes needed to transform the current food systems into practices that regenerate the soil, watersheds, food quality and economies, allow biodiversity to flourish and mitigate climate change. ■

Contents

Acknowledgements

I would like to thank my very lovely husband Mark for supporting me on the journey of creating this book – the farm walks, the endless discussions, the lack of company for very many weeks and days while I tried to capture in words what I know from the fields. I would also like to thank my beautiful daughters Ruby and Lily-mei who have joined us on many farm walks, listened politely to so many discussions, enjoyed the food at the table and share my love of a good farmers' market.

Thank you to Gabriel Kaye and Martin Large, who as the Biodynamic Land Trust's directors gave their trust to and encouraged the embryonic Apricot Centre team and helped shape Huxhams Cross Farm into what it is today. Thank you to Anne Phillips and Wendy Cook who were key in bringing Huxhams Cross and the Apricot Centre together. And thank you to the 150 shareholders who invested in the farm. Martin Large also encouraged me to write this book so that some of the knowledge we gained can be shared with others.

The late Professor Martin and Ann Wolfe of Wakelyns Farm laid the foundations for the very core of the premise behind this book. I would like to thank them for all the farm walks, talks, dinners, laughter, thoughts, challenges, the occasional raised eyebrow and the very many flapjacks – but also the encouragement over 20 years to keep going with this work.

Thank you to all of the farmers and growers who gave me permission to use them in case studies to showcase different aspects of the food systems. Most of them use multiple methods but I have generally illustrated just one aspect of their work. They include the farmers at Winter Green Farm, Oregon; Christine Arlt at Sekem, Egypt; Tamarisk Farm; Sarah Green of Mark Farm; the Organic Lea team, London; la Ferme du Bec Hellouin, France; Mark Lea of Green Acres Farm, Shropshire; Emmanuel Baya of Margarini Children's Centre and Organic Demonstration Farm, Kenya; David Wolfe, the late Martin and Ann Wolfe of Wakelyns Farm, Suffolk; Martin Crawford of the Agroforestry Research Trust, Devon; Jon and Lynne Perkin of Parsonage Farm, Dartington: Josiah Meldrum of Hodmedods; Jon Perkin, Dan Mifsud and Bob

Mehew of Dartington Mill; and Julie Brown and Kerry Rankine of Growing Communities, London.

Thank you to those who took the time to contribute to the book: Julia Wright and Luke Owen of the Centre for Agroecology and Water Resources at Coventry University, Philip Franses and Mini Jain of the Flow Partnership, Rachel Bohlen of the Apricot Centre, Christian Kay for his wonderful photos, Ben Raskin for his agroforestry photos, Sophy Banks for insight into ecopsychology, and Christopher Upton of Zerodig at Oakbrook Community Farm.

Thank you to the Hawthorn Press team, Anthony Nanson for the huge amount of editing required and Amanda Cuthbert, my neighbour, who helped me structure my thoughts early on in the process.

Thank you to the Devon Environmental Foundation for its generous support of our research, together with its impact assessments and its encouragement to bring these messages into the world.

Last but not least, thank you to my wonderful colleagues at the Apricot Centre.

It has been a huge team effort in which everyone has brought their strengths and good humour to make Huxhams Cross Farm work so well, including now to offer it as a demonstration farm for others wanting to make the regenerative transition. They include our directors, in particular Matt Harvey, who has kept us laughing no matter what, and Rhodri Samuel and Liliene Uwimame; our Friends of the Apricot Centre, who have guided us and taken time to input into its development – Wendy Cook, Anne Phillips, Anne Marie Mayer and George Sobel; the operational team – Mark O'Connell, Rachel Phillips, Bob Mehew, Dave Wright, Caspar Meridith, Ross Perret, Rachel Bohlen, Amy Worth, Richard Andrewes and Keren Kossow; the whole well-being team of therapists and mentors doing incredible work in and around the farm; apprentices past and present; all the volunteers who help us; and, of course, our customers who enjoy our food.

I would like also to offer gratitude to the tiny piece of land that we call Huxhams Cross Farm, for its bounty and the joy that it gives us all. ■

Foreword

When you come home from the daring journey, the demons slain and the elixir cradled in your palm, what do you find? The old world is still indifferent. It still does not know it needs your magic. Now the work begins. That was the old way, now we must do this thing. 'To make an end is to make a beginning', wrote T. S. Eliot in *The Four Quartets*. 'The end is where we start from.'

And what a place to start from, at the Earth's great interlocking crises. The loss of biodiversity and species, the crushing of the climate, the rise of inequality, the loss of contentment, the relentless pursuit of material consumption. In the modern world of affluence, many things have been getting better, but some suddenly became much worse. Once upon a time, we knew what a good agriculture and food system could look like, and yet somehow it slipped from our grasp. We might well ask, again, how might greener, low-carbon and healthier options emerge?

When you enter the forest at its darkest point, wrote Joseph Campbell, there is no path. If you find one, it is probably someone else's. The idea is to make your own way. It's over there, the start line. We just need to get in the game, to gather up a staff and enough food and possessions. And start walking.

Well, Mary Oliver had a marvellous answer in her wonder-poem called *Sometimes*:

Instructions for Living a Life:
Pay attention.
Be astonished.
Tell about it.

And this brings us to Huxhams Cross Farm, called a few short years ago by a local rural contractor 'a miserable bit of land'. The world needs transforming; it needs leadership. Someone needs to walk the path over each piece of such land. And this, we see, is what Marina O'Connell has done with glory in this powerful and personal book about transforming the land for the better. Marina O'Connell weaves inspirational stories of redesign and transformation, showing how regenerative methods for agriculture and food have come to life. In half a decade, she created a productive, diverse, profitable and regenerative farm from depleted soil, and has said, 'Over here is a path, now we can walk it'.

The concern for sustainability in agroecosystems centres on the fundamental importance of both agricultural and nonagricultural ecosystems and their links with farmers and consumers. Agriculture is unique as an economic sector as it directly affects many of the very natural and social assets on which it relies for success. These influences can be both good and bad. Industrialised and high-input agricultural systems rely for their productivity on simplifying agroecosystems, bringing in external inputs to augment or substitute for natural ecosystem functions, and externalising costs and impacts. Pests tend to be dealt with by the application of synthetic and fossil-fuel-derived compounds, wastes flow out of farms to water supplies, and nutrients leach to the soil and groundwater. As a result, there has been widespread and increasing cost to natural ecosystems and human health.

By contrast, regenerative approaches to agriculture seek to use ecosystem services without significantly trading off desired productivity. When successful, the resulting agroecosystems have a positive impact on natural, social and human capital, while unsustainable ones continue to deplete these capital assets. A wide range of different terms for more sustainable agriculture have come into use: regenerative agriculture, a doubly green revolution, alternative agriculture, an evergreen revolution, agroecological intensification, green food systems, save and grow agriculture,

and sustainable intensification. Many of these draw on earlier traditions and innovations in permaculture, natural farming, the one-straw revolution, and forms of biodynamic and organic agriculture.

We now know that the concept of sustainability should be open, emphasising values and outcomes rather than means, applying to any size of enterprise, and not predetermining technologies, production type, or particular design components. Central to the concept of all types of regenerative systems is an acceptance that there will be no perfect end point due to the multi-objective nature of sustainability. Thus, no system is expected to succeed forever, with no package of practices fitting the shifting ecological and social dynamics of every location. In the 1980s, Stuart Hill proposed three non-linear stages in these transitions towards sustainability: i) efficiency; ii) substitution; and iii) redesign. While both efficiency and substitution are valuable stages towards system sustainability, they rarely achieve the greatest co-production of both favourable agricultural and environmental outcomes at regional and continental scales.

In the first stage, **efficiency** focuses on making better use of on-farm and imported resources within existing system configurations. In the second stage, **substitution** focuses on the replacement of technologies and practices. The third stage incorporates agroecological processes to achieve impact at scale; **redesign** centres on the composition and structure of agroecosystems to deliver sustainability across all dimensions to facilitate food, fibre and fuel production at increased rates. Redesign harnesses predation, parasitism, allelopathy, herbivory, nitrogen fixation, pollination, trophic dependencies and other agroecological processes to develop components that deliver beneficial services for the production of crops and livestock. A prime aim is to influence the impacts of agroecosystem management on externalities (negative and positive), such as greenhouse gas emissions, clean water, carbon sequestration, biodiversity, and dispersal of pests, pathogens and weeds. While efficiency and substitution tend to be additive and incremental within current production systems, redesign brings the most transformative changes across systems.

But for redesigned agricultural and landscape systems to have a transformative impact on whole landscapes, this requires cooperation, or at least individual actions that collectively result in additive or synergistic benefits. For farmers to be able to adapt their agroecosystems in the face of stresses, they will need to have the confidence to innovate. As ecological, climatic and economic conditions change, and as knowledge evolves, so must the capacity of farmers and communities to allow them to drive transitions through processes of collective social learning. This suggests redesigned systems have the valued property of intrinsic adaptability, whereby interventions that can be adapted by users to evolve with changing environmental, economic and social conditions are likely to be more sustainable than those requiring a rigid set of conditions to function. Every example of successful redesign at scale has involved the prior building of social capital, in which emphasis is placed on relations of trust, reciprocity and exchange, common rules, norms and sanctions, and connectedness in groups. As social capital lowers the costs of working together, it facilitates cooperation, and people have the confidence to invest in collective activities, knowing that others will do so too.

All things are connected. And this is why land and agricultural transformations such as these described in this timely book on designing regenerative food systems are so important. Can we do better, if we think differently? The answer is a resounding yes. The next question then centres on what could happen next. Regenerative agriculture approaches have been shown to increase productivity, raise system diversity, reduce farmer costs, reduce negative externalities, and improve ecosystem services. There are thus a range of potential motivations for farmers to adopt agroecological approaches on farms, and for policy support to be provided by national government, third sector and international organisations. But these transitions still require investments to build natural, social and human capital: redesign is not costless.

There are important arguments that suggest the world would not need to increase agricultural production if less food were wasted, and less energetically-inefficient meat was consumed by the affluent. These changes would help, but there is no magic wand of redistribution. Most if not all farmers need to raise yields while improving environmental services. And now we know, these changes are happening worldwide. Two groups of 40 authors have recently undertaken global assessments of the spread of these sustainable practices: 160 million farms, 450 million hectares and 240 million people organised into social groups to take action at the landscape level.

It was the questions from visitors about the transformation of Huxhams Cross Farm's depleted soil and bare land that sparked this book. They wanted to understand what they saw so that they could go back and redesign their own farms and communities. This evidence shows that redesign of agro-ecosystems around agroecological and regenerative approaches to sustainability can achieve yield increases. The evidence on farms of redesign and regenerative transformations offers scope for optimism. The concept and practice embodied in the application of agroecology will be a process of adaptation and redesign, driven by a wide range of actors cooperating in new agricultural knowledge commons and economies.

Jules Pretty
Professor of Environment and Society,
University of Essex

Bibliography

Blamires H, 1969, *Word Unheard: A Guide through Eliot's Four Quartets*, Methuen, London.

Campbell J, 1949 (2008), *The Hero with a Thousand Faces*, New World Library, Novato, California.

Hill S, 1985, 'Redesigning the Food System for Sustainability', *Alternatives* 12, 32–36.

Pretty J, 1995, *Regenerating Agriculture*, Earthscan, London.

Pretty J, 2003, *Agri-Culture*, Earthscan, London.

Pretty J, Benton T G, Bharucha Z P, Dicks L, and 11 more authors, 2018, 'Global Assessment of Agricultural System Redesign for Sustainable Intensification', *Nature Sustainability* 1, 441–446.

Pretty J, Attwood S, Bawden R, van den Berg H and 25 more authors, 2020, 'Assessment of the Growth in Social Groups for Sustainable Agriculture and Land Management', *Global Sustainability*, 3 e23, 1–16.

Introduction

In order to create the sustainable farms now urgently needed for the 21st century it is useful to have available a 'toolkit' of methods by which to radically transform a piece of land, or at least to nudge food production in a more sustainable direction. All the methods described in this book can be used in a pure form by themselves. Each system appeals to individuals and communities in different ways. However, in my experience and from a farming perspective, these various methods weave together to create resilient, low-carbon and productive biodiverse farming systems. They contribute to what I have called 'the sustainable agricultural revolution'.

Visitors and students on courses at Huxhams Cross Farm have asked me how we created a productive, beautiful, profitable and regenerative farm from depleted soil on former land of Dartington Estate. The contractor who had previously worked this land had called it 'a miserable bit of land'. The short answer to the question is that we drew on the methods described in this book to create a sustainable farm from industrially farmed land. We observed that many visitors understood one sustainable farming system but rarely grasped the variety of approaches and how to weave them together. The methods are culturally different, but from a farmer's perspective they complement each other very well, each bringing different strengths.

Relatively few people fully understand what biodynamic farming, organic farming, permaculture, agroforestry, agroecology and regenerative agriculture are, how they relate to each other and how they compare with current industrial farming models. This book aims to give an overview and insight into these systems from a practitioner's perspective. It does not provide an in-depth academic study of any of these systems. At the end of each chapter are signposts to further sources of information to explore these systems – books, websites, films, academic papers, and real or virtual farm visits. Each system is illustrated by an existing case study of a farm working in the ways described. The case study of Huxhams Cross Farm showcases how the systems can be brought together to transform land in a short period of time.

Although each chapter can stand alone, the structure of the book reflects how food systems and farming require an integrated holistic systems approach, rather than a fragmented reductionist approach. There are many overlaps between the chapters, just as there are many overlaps between farming and food systems. What is good for the soil biome is good for plant nutrition, is good for biodiversity, is good for human health and is good for the economic health of a farm. Each farming system described here brings a different quality to a comprehensive holistic systems approach to sustainable farming and food systems.

My farming story

I have been professionally involved in sustainable farming and growing since the 1980s. I started with a degree in horticulture from the University of Bath, where I was trained in the industrial methods of the day. I came across my first biodynamic farms in Ireland and Brazil by accident during my work experience placements and was amazed by the quality of the food, the physical beauty of the farms, and the pleasant nature of the work in comparison with working in industrial farming systems. On leaving university, my first job was at the Horticultural Training Workshop at Dartington Hall Trust, South Devon, training young people to become gardeners or nursery workers. I was privileged to attend one of the first permaculture design courses in the UK, and I attended weekly study groups for biodynamic farming at the same time. One of my first 'aha!' moments was at a talk by Peter Procter, one of the world's great biodynamic trainers, at the local Steiner school. He drew a picture of a biodynamic farm that was identical to the permaculture notion of a 'zoned' farm. It was then that I realised that there are more similarities between the two systems than most people thought.

I hung out with pioneering organic and biodynamic growers and farmers in the area, often volunteering to help out at the weekends. I attended sustainability conferences. When Schumacher College opened in 1991, I went to evening talks by some of the leading thinkers in the field of sustainability. I have wonderful memories of delivering piles of fresh herbs, vegetables and fruit into the college kitchen where Julia Ponsonby cooked them up into delicious food, and of Thursday evening fireside talks by the likes of Satish Kumar, Wendell Berry and Vandana Shiva. My ways of thinking about growing food changed from how I had been taught in university to new ways that were and still are evolving through practice. Such learning by doing is an example of what is sometimes called 'action learning'.

I first used permaculture design methods on a large scale to develop the Organic Market Garden at Dartington Hall between 1989 and 1991. This evolved over the next 30 years into the successful School Farm CSA (community-supported agriculture) that it is today. My partner, Mark, was studying psychotherapy at this time and we were both reading the same books, which intrigued me. What was the crossover between sustainable land practice and healing people from trauma? It is only now, 30 years later, that we have finally been able to bring these two disciplines together fully in practice and understand how they interweave.

We moved to East Anglia, where I worked at Otley College of Agriculture, lecturing in permaculture, biodynamic horticulture and organic farming, and offering continuing professional development (CPD) to help existing farmers change their practice. The need for this very enlightened programme was not fully recognised at the time and so the funding for it was removed. During this time I also managed to squeeze in a master's degree at the University of Essex. This transdisciplinary degree in environment and society offered modules in environmental politics and sociology, giving me further insight into how sustainable systems of food production did and did not work. The degree course was led by Jules Pretty, one of the leading lights in sustainable agriculture. We met Martin and Ann Wolfe at Wakelyns Farm in 1996 and took groups of students and farmers to visit the new agroforestry farm. Over the years, I observed how the agroforestry system developed into what it is today. We met up regularly with Ann and Martin to discuss the more holistic thinking that was generating and integrating the new sustainable food systems practices.

Leaving the college work, Mark and I managed to buy a 1.5-hectare field with a home. We used permaculture design methods to create a beautiful orchard, combined it with agroforestry throughout the site and applied biodynamic preparations regularly on the depleted soil. We employed a wonderful farm worker called Wayne for two days a week. When we left, the farm was turning over £25,000 per year. The holding was registered organic. We brought up our children there and built the first Apricot Centre out of recycled shipping containers. We sold the produce at the renowned Growing Communities Farmers' Market in Stoke Newington, part of a network of sustainable food producers in and around London.

We both worked elsewhere as well to make ends meet. Mark worked for the National Health Service (NHS) as a child psychotherapist, in particular with children either adopted or in the care system, having suffered some trauma in their lives. I worked as a creative practitioner in and around schools in deprived areas of Essex, collaborating with the teachers to create

outdoor classrooms and deliver the curriculum in a kinesthetic learning environment, using the activities of growing, cooking and eating food to teach the children maths, science and English. We were also a part of the Transition Towns movement in East Anglia, attempting to create local food systems in the region. I delivered some Transition Town training in London and East Anglia.

What I've described in a few sentences encompasses 16 years of work. As I observed and worked the small farm, and it matured from an ordinary flat oblong of depleted pasture, something magical happened. It became very abundant. The food tasted delicious. The place was and still is full of wildlife. We had sparrowhawks, turtle doves, grass snakes, owls, foxes and elephant hawkmoths – the dawn chorus was a racket. It was then that I really knew how well these farming systems work.

In 2014, Martin Large, founding director of the Biodynamic Land Trust, asked us to submit a business plan for leasing their new 13-hectare bare field site on the edge of Dartington Hall Estate. The Biodynamic Land Trust intended to buy the piece of land and needed a farmer and team to transform it into a financially independent biodynamic farm. Bob Mehew joined us as a director with skills in project management and financial planning. The farm was bought in 2015 and is now a fully operational biodynamic farm designed using the permaculture design process. Agroforestry is woven through the site, which is home to a thriving well-being service for young people and families.

Rachel Phillips joined the well-being team to undertake the complex task of bridging the well-being work and the farm; she created a plethora

Figures 0.1 and 0.2: Transformation of the soil at Huxhams Cross Farm – left: before and after.

of nature-based activities for young people. Dave Wright joined the farm team to manage the market garden, which now yields delicious vegetables, chickens, eggs and wheat. The soil has been transformed after five years of sustainable farming practices following 40 years of industrial barley crops. From a standing start, the annual turnover after five years is £200,000+ for the farm and another £400,000 for the well-being service. We employ six people on the farm, with three apprentices per year, and another five people in the well-being team, as well as a number of part-time psychotherapists and psychologists. The farm is now sequestering ten tonnes of carbon dioxide equivalent per year – twice as much as the farm uses. The biodiversity is up, people who eat our food say it's delicious and most of them say they eat more fruit and vegetables because of it. We deliver almost 4,000 hours of

therapy per year in total, including 500 hours on the farm. Over 1,000 people per year attend training or well-being activities such as 'mud tots', a parent and toddler group in the forest garden area. The farm attracts cohorts of young people as apprentices who are keen to learn about sustainable forms of food production and bring their own skills and knowledge to the the farm, the business or the Apricot Centre. All of this has come about from the transformation of the soil (Figures 0.1 and 0.2).

I am descended from seven generations of nurserymen from the small village of Boskoop in the Netherlands. Famous for its fruit tree 'Belle of Boskoop', Boskoop is central to the nursery stock production of trees in Europe. My generation is the first generation of my family in which women too have worked professionally in horticulture. During my career I have found myself surrounded by more and more wonderful women pioneering the shift to sustainable food production and relocalised food systems.

I grew up listening to my mother's stories about the Winter of Hunger in the Netherlands. She was 18 in 1944 and she suffered from malnutrition at this time. She told me how she gleaned the fields with her brother for potatoes and peas, and of her relatives walking for hours to share some of this food. She spoke of seeing people starving to death in the streets. My English father told me stories of his childhood during the Great Depression, in 1935, when he was ten, there was no food in the cupboard when he came home from school. When I reflect on all this it seems no surprise that my whole career has revolved around growing food.

I would like to acknowledge the late Martin and Ann Wolfe, who have had a huge influence on my thoughts and the approach presented in this book. Their thinking about sustainable food production systems has resonated with me since we met: the need to go back to the point of divergence of the industrial and the sustainable models of food production and to devise modern sustainable systems rather than small adjustments to the current industrial farming system. I have had the privilege to visit Wakelyns Farm from 1996 until the present day and have observed it develop – and eaten many of Ann's delicious flapjacks in the process.

Overview of the book

Part 1 of this book's three parts establishes the context. Chapter 1 outlines the extent of the challenges facing food systems in the 21st century. Chapter 2 tells the story of how we got to this point and explains the structure of the book. Since the advent of industrial food systems, there has been a parallel development of diverse sustainable food systems. These have been quietly developing their practice and principles, often underrated, misunderstood or simply ignored.

Part 2 comprises six chapters, one on each of the sustainable food systems that have arisen sequentially from that point of divergence between industrial and sustainable food systems. These are the biodynamic, organic, permaculture, agroforestry, agroecology and regenerative food systems. Each chapter explains what the respective system is, where it started, its principles and practices, why it works, what it looks like in terms of one or two case studies, and where training is available.

Part 3 looks at how these sustainable food systems can be used as a toolkit to revolutionise the food systems of the 21st century. Chapter 9 highlights the characteristics that will be required of sustainable farms if we are to meet the challenges of climate change, biodiversity loss and producing enough food. The chapter then illustrates, with the use of research findings and practices arising from our six sustainable methodologies, how these food systems have pioneered holistic solutions to these challenges.

Chapter 10's case study of Huxhams Cross Farm illustrates how, in a few short years, the weaving together of diverse sustainable farming practices transformed a barren piece of land into the thriving healthy farm that it is today. Chapter 11 explores the next steps in the transition to sustainable food systems and how everyone can choose to be part of the sustainable agricultural revolution. ∎

Part 1

the challenges facing food production

and how we can meet them

Chapter 1

The challenges facing food production

Introduction

Farming – and, in particular, sustainable farming – is a slow business. It takes time to see what effects change in a food production system and to understand with confidence how such a system works and how it can be replicated. The farming methods described in this book could be likened to a fringe arts festival – only it's a festival that's been happening in slow motion for the last 100 years. Fringe festivals are exciting spaces where the cutting edge of experimental arts may be found. They are spaces for mistakes, for the avant-garde, for creative new directions that then seep into the mainstream and become the norm until the next generation of creative people come through. It is on the edges, in marginal areas, that diversity is at its greatest, where new ideas flourish and the improbable can happen.

In ecology, it is similarly on the edges of ecosystems that the greatest diversity of species exists. It is at the edges that ecological resilience is greatest and the species are found that are most resilient to fast-changing circumstances.

We are used to thinking about the fast-moving art scene in this way. In contemporary food production too we can see a 'fringe'

movement, only it's slow moving as all farming and gardening are by nature. This book explores the systems found at the edges of the world of food production. It seeks to uncover how they can provide us with answers to some of the huge challenges that will face food production over the next 50 years. Food production is at a pivotal moment. In this time of escalating climate crisis, biodiversity loss and expanding population, creative and radical approaches are needed to pioneer ways we can feed ourselves with quality, healthy food without destroying the very biodiversity and environmental services upon which we depend. Because of the time it takes to develop new sustainable farming systems, the time to act is now. More resilient food production systems, farms and farmers are urgently needed. Climate change will not wait for us to catch up. The systems described in this book have been proven to work. 'Whatever you can do or think you can, begin it. Boldness has genius, power and magic in it.'[1]

The challenges facing food production

This book aims to be positive and solution based. However, to seek the right solutions we first need to understand the challenges that face food production today. This can be overwhelming and depressing, for they are world-scale challenges. The aim of Part 1 is to introduce these challenges, in the context of food systems, in order to understand better the solutions that will be suggested.

The four main challenges to food production in the 21st century constitute a perfect storm: the need to both mitigate and adapt to climate change; catastrophic loss of biodiversity; the rising human population; and rising health issues owing to poor diet in the global north and food shortages in the global south. In other words, we need food production systems that sequester carbon, reduce fossil fuel use, are resilient to extreme weather, support biodiversity and are productive enough to feed the growing population with healthy food.

While doing this the food producers must be able to earn enough to support their own livelihood.

This book looks to the edges, where people are passionate, creative and radical in their response to these challenges. It is here that we find some of the very few ways in which all four of the challenges intersect and can also be met. The toolkit of sustainable food production methods addresses all four challenges simultaneously. All of the systems outlined have arisen from an ecological paradigm of food systems which is divergent from the industrial productionist paradigm that is predominant today, sometimes called 'conventional agriculture'. They amount to an ongoing radical rethink of food production systems. The four challenges and our understanding of them are changing all the time, but their core elements remain the same.

Challenge 1. Climate change mitigation

Industrial food production systems need to reduce the amount of carbon they use to net zero. Farms uniquely have the ability to sequester carbon in their soils and trees. They are able not only to decarbonise themselves but also to mitigate climate change by soaking up carbon produced from other sectors.

Current industrial food production systems contribute approximately 30 per cent of carbon emissions. That comprises approximately 10 per cent from food production as far as the farm gate, and a further 20 per cent from distribution and packaging.[2] There are four main greenhouse gases (GHGs): carbon dioxide, methane, nitrous oxide and fluorinated gases. In farming, our main focus is on the first three GHGs.[3]

Farms use fossil fuels, and burning fossil fuels produces carbon dioxide. They are used to power tractors and machinery and to generate electricity for cold stores and drying and processing equipment. Fossil fuels are used for shipping, airfreight and refrigerated trucks to transport food over long distances around the world in different seasons. These distances are known as 'food miles'. Out-of-season food is produced in temperate climates in glasshouses using energy from fossil fuels. The 'embedded energy' in a product is the amount of energy required to make it. Fertilisers, pesticides and plastic all entail high embedded energy.

Nitrogen fertilisers and animal manure emit nitrous oxide, a GHG 300 times more powerful than carbon dioxide. Nitrogen-based fertilisers are also thought to contribute six per cent of worldwide carbon emissions.[4]

Cattle produce methane, which is 30 times

as powerful a GHG as carbon dioxide.

Many cattle are grass-fed on the grasslands that occupy 70 per cent of the world's farmland. However, increasingly cattle are reared for beef in 'feedlots' or giant sheds. They are fed grain or soya that is shipped from all over the world and produced on a large scale using tractors, pesticides and fertilisers, often on land that is cleared rainforest. A Chatham House study showed that livestock production is responsible for 15 per cent of worldwide carbon emissions, equivalent to the exhaust gases from vehicles[5] or half of the world's carbon emissions from food production as a whole. Meat consumption is increasing; in the global north it is two to three times what it should be for good health, and it is also increasing in the global south.[6] However, holistic grazing techniques (described in Chapter 8) can help pasture-fed cattle to sequester carbon in the soil and repair soils after arable crops and in grasslands suffering from desertification.[7]

Of the GHG emissions from farming in the UK, 70 per cent of the greenhouse impact comes from nitrous oxide emissions, 49 per cent from methane emissions and 1.6 per cent from carbon dioxide emissions.[8]

Not only do fossil fuels contribute to climate change, but they are also a finite resource. Peak production of oil occurred somewhere between 2010 and 2015, which means that oil will become more difficult and expensive to access. Extraction of oil from sources ever more difficult to access, using fracking, tar oil extraction, deep sea drilling and long pipelines, is doing increasing harm to the environment. As oil becomes increasingly scarce it is drilled from new places and in ever more complex situations, such as

the deep sea. As the Arctic ice cap melts, many countries and companies plan to drill for oil in this challenging region. Thus accidents will become more likely and there will be increasing risks of oil spills and environmental damage.[9]

Although fossil fuels are abundant, and oil is not currently in scarce supply, there will be less in the future. Oil's increasing scarcity and price volatility causes conflict and economic turmoil. Food production systems that are independent of oil and oil prices provide greater resilience in the price and availability of food.

The volatility of oil prices has led governments and oil companies to seek alternatives to fossil fuels for transportation. Hence the development of biofuels based on ethanol made from sugar cane and rapeseed, sunflower and palm oils. The land required to grow these fuel crops is vast and comes at the expense of using that land for food production.

With suitable management, farms can sequester carbon in their soils and trees. With an increase in soil organic matter (SOM) there is an increase in carbon sequestration, which enables farming not only to decarbonise its activities but to become carbon negative, that is, to sequester more carbon than it uses, so mitigating climate change. This is a complex field whose science is developing all the time (see Chapter 9).

The International Panel on Climate Change (IPCC) set target carbon emission reductions at the Paris Conference of Parties (COP) in 2015. This was ratified by most of the world's countries. The current aim is to reduce emissions by 100 per cent by 2050, in order to limit the global temperature increase to a rise of

Global greenhouse gas emissions from food production

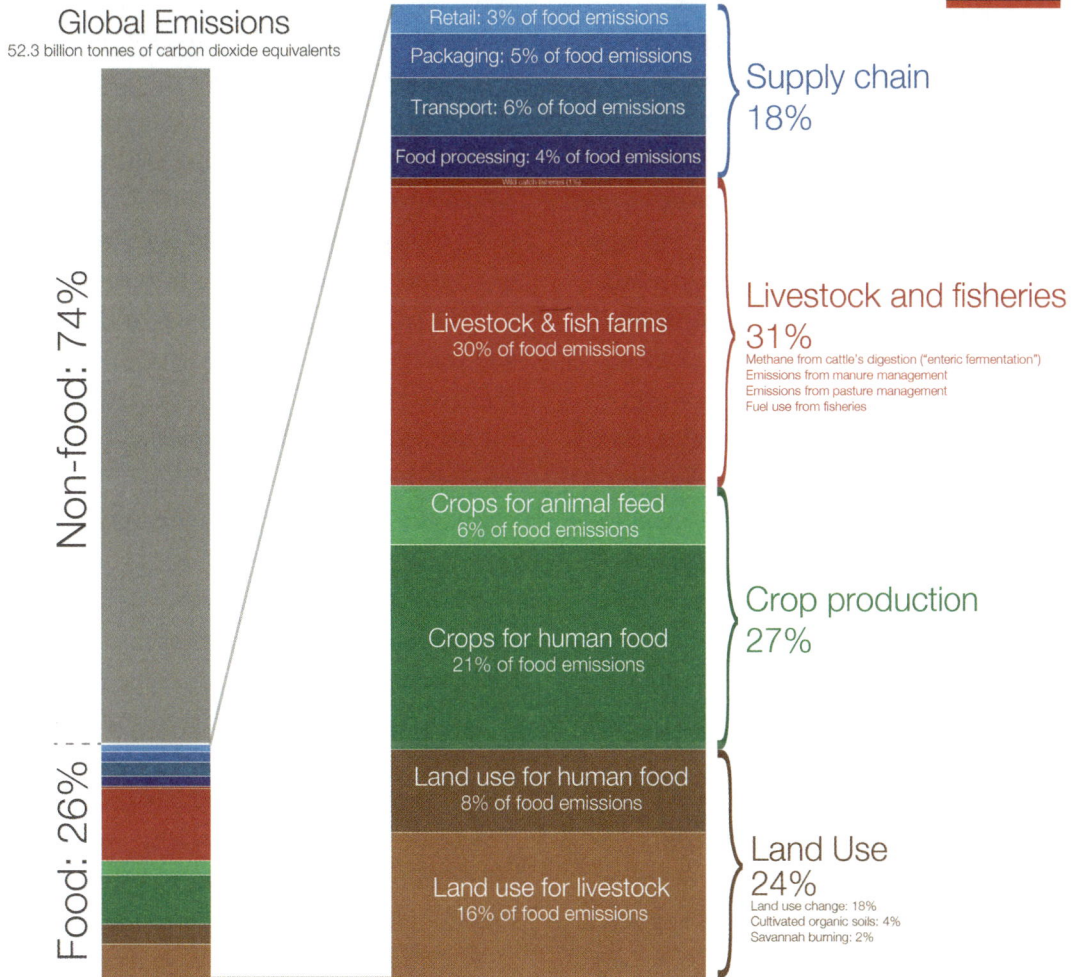

Our World in Data

Global Emissions
52.3 billion tonnes of carbon dioxide equivalents

Non-food: 74%

Food: 26%

Retail: 3% of food emissions

Packaging: 5% of food emissions

Transport: 6% of food emissions

Food processing: 4% of food emissions

Supply chain 18%

Livestock & fish farms
30% of food emissions

Livestock and fisheries 31%
Methane from cattle's digestion ("enteric fermentation")
Emissions from manure management
Emissions from pasture management
Fuel use from fisheries

Crops for animal feed
6% of food emissions

Crops for human food
21% of food emissions

Crop production 27%

Land use for human food
8% of food emissions

Land use for livestock
16% of food emissions

Land Use 24%
Land use change: 18%
Cultivated organic soils: 4%
Savannah burning: 2%

Figure 1.1: Carbon emissions from agriculture.

Data source: Joseph Poore & Thomas Nemeck (2018). Reducing food's environmental impacts through producers and consumers. Published in Science. *OurWorldinData.org – Research and data to make progress against the world's largest problems. Licensed under CC-BY by the author Hannah Ritchie.*

1.5°C.[10] Some countries are now aiming for 100 per cent emission reduction by 2030 because of the recently declared climate change emergency. Change in farming practice needs to start now if it is to be effective in reducing emissions by 2030. In the UK the National Farmers' Union (NFU) has set the farming sector a target of zero net carbon emissions by 2040. As changes in farming are slow by nature, the necessary changes are now urgent if they are to be effective. The time to start creating 'farms for the future' is right now.

Challenge 2. Climate change adaptation

Climate change produces erratic weather. Predictions of changes in weather patterns suggest there will be more extreme weather events such as storms, high rainfall, high wind, rain at the wrong time of the year, drought, late frosts and warm spells in winter. Climate change is now experienced in real time and is no longer something projected into the future. Farming by its very nature relies on predictable weather patterns. One late frost can wipe out a fruit or vegetable harvest. A drought in summer can cause crop failures. Storms devastate glasshouses and polytunnels. Wet autumns make autumn cereals difficult to sow and harvest. Flooding is becoming a regular occurrence. Farms are a part of ecosystems. They can also be a part of the solution in, for example, regenerating watersheds and flood alleviation.[11] Farms of the future will need to be reslient and to adapt to climate change.

Challenge 3. Offsetting biodiversity loss

Worldwide biodiversity loss is currently estimated, by the United Nations Environment Programme (UNEP), to be 1,000–10,000 times the background rate of extinction. The term 'biodiversity' encompasses the diversity of species in a habitat, the number of habitats or ecosystems present in a region or farm, and the genetic diversity within each species. A broad range of biodiversity gives ecosystems the ability to adapt to climate change. Biodiversity includes, besides plants and animals, the microorganisms in the soil.[12]

'Agro-biodiversity' refers to the number of crop plant species and varieties and animal breeds used in food production. The diversity of all of these is in decline: the loss of agrobiodiversity has been estimated as 75–90 per cent over the last 100 years. This has made people's diet less diverse and cropping systems less resilient. New varieties able to cope with climate change are bred from the genetic banks of crop plants. If these are lost then so is the ability to breed new crops. Older varieties of crops are often more 'nutrient dense' and more nutritious.[13]

The causes of losses in biodiversity are wide-ranging, but one of the main drivers is the loss of habitat to agriculture. This includes the loss of forest, hedgerow and woodlands and the draining of wetland meadows, filling in of ponds and ploughing of marginal land to serve the mechanisation of food production. The use of fertilisers and pesticides causes the loss of microorganisms and organic matter in the agricultural soils. These are the foundation of the higher food chain. If there are fewer fungi and bacteria, there will be fewer worms, insects, small mammals and birds. The ensuing fragmentation and loss of wildlife corridors in the remaining habitat results in areas of habitat too small to support self-sustaining populations of wildlife. Pollution further degrades habitats and so puts its wildlife under further stress. Climate change will exacerbate this problem further, since species will need to migrate if they are to adapt to climate change, in order to keep finding suitable conditions in which they can live, feed and breed. Lack of wildlife corridors makes this extremely difficult.

Since the introduction of industrial farming systems after World War II, biodiversity levels have plummeted. Agricultural soils are so degraded around the world that the

Why is biodiversity important?

Biodiversity has intrinsic value. A species of moth may not have any specific use or monetary value to people, but it has value in its own right, as a moth. From an 'ecocentric' world view, human beings are part of biodiversity not separate from it. Most indigenous food systems locate people as part of the web of life. So many people love biodiversity. We feed the birds in the back garden. We join the Royal Society for the Protection of Birds (RSPB) by the million. We pay money to 'Save the Tiger' campaigns even though we shall probably never see a tiger in the wild. We just like to know that 'wildness' is there somewhere. So biodiversity is important to us psychologically.

Biodiversity also provides 'environmental services'. Woodland and forests soak up carbon emissions, clean the air and drive climate systems. Estuaries with complex reed beds filter polluted water draining from agricultural land, where nitrate fertilisers leech from the soil, and so clean it up before it reaches the sea. Agricultural soils contain fungi, bacteria and invertebrates. These organisms feed on organic matter and form the bottom of the food chain for farmland birds and mammals. The organic content of the soil provides nutrients for crops, increases water-holding capacity and decreases erosion. Without these systems our planet's biosphere would not function.[18]

United Nations (UN) suggests there are only 60 harvests left in some soils. In the UK, the Department for Environment, Food and Rural Affairs (DEFRA) suggests that some soils have only 40 harvests left.[14]

The concept of 'land sparing' proposes that if some land is farmed more intensely then other land can be 'spared for nature'. However, this does not address the systemic problems in the long term. Firstly, it does not address the problem of habitat fragmentation. Secondly, the soils that would be used continually for industrial food production will continue to decline and will only last another 40–60 years without regeneration. Therefore land sparing only postpones the problem, rather than solving it. Using more pesticides and fertilisers will only exacerbate climate change and the loss of biodiversity.[15]

In the UK and much of continental Europe, where farmland occupies 70 per cent or more of the land mass, it is crucial that farming systems support biodiversity at all levels. Ideally, farmland will be 'porous' to biodiversity and the wild organisms will be able to move from farm to farm via wildlife corridors.

In Africa, Indonesia and Amazonia the same issues are in play. Clear felling rainforests for the production of palm oil, biomass fuels, soya, or grain to feed beef cattle, or for cattle ranching, squeezes out native species, some of which have not yet been identified.

Single-use plastic packaging is used to wrap food, especially food sold by supermarkets. Very few plastics are recycled. The majority are thrown away and end up in landfill or incinerators or find their way into rivers and oceans, where the currents move them into giant clumps of floating plastic. The 'Great Pacific Garbage Patch' is made up of 1.6 million km^2 of plastic rubbish, which slowly degrades into microplastic beads that are then eaten by fish. This causes malnutrition, since plastic is not food. Larger plastic bags, nylon fishing nets and other pieces of plastic damage marine animals such as turtles who mistake them for jellyfish. Animals also become entangled in such things. Plastics contain bioaccumulative toxins that move up through the food chain to people.[16] Plastics used to wrap food and line food cans contain chemicals that mimic hormones and are thought to be contributing to dramatic falls in fertility in both people and animals.[17]

Challenge 4. Producing enough food for a growing population

Population growth is the elephant in the room in many discussions about global food supplies. It is the overall population in the world that is important. Although the population of the global north is projected to decline, the population of the global south is increasing. It is predicted that the world's total population will peak in 2050 at around 10 billion, before then beginning to decline. To feed this population, the Food and Agriculture Organization (FAO) estimates, food production will need to increase by 60 per cent.[19]

It is estimated that a third of all food is currently wasted through long supply chains.[20] Land that was used for food production in the past is now underutilised because it has been degraded, or because farming is no longer an economically viable way to make a living. The majority (40 per cent) of grain now grown in the world goes to feed livestock and to produce biofuels although a proportion of this would be inedible to humans.[21]

There is no more new land in the world that can be brought into food production, since the land that is left is required for biodiversity and environmental services. Farmland is lost to the building of houses and urbanisation and has been degraded by years of farming with pesticides and nitrogen fertilisers, which have led to soils with low organic content.

There is also the issue of 'food and health'. The question is not just 'Is there enough food to feed the increasing population?' but 'Can people be fed with healthy food?' Food policy discussions tend to focus on scarcity of food rather than the potential abundance of it, and are based on quantity rather than quality, and fear rather than equity. The term 'nutrition transition' was coined to describe a phenomenon that started in the 1970s in the US and has since been observed across the globe as countries replace local food cultures with industrial farming and agribusiness. Displaced peasants and farmers are moved into the cities. The countryside empties as farms get bigger and more industrialised. The 'nutrition transition' involves a shift toward a diet with more processed foods, meat, sugary drinks and high-fat snacks, much of this being wrapped in plastic. This change in diet causes health issues such as obesity, hypertension, diabetes and heart disease. These illnesses, pre-COVID-19, have been the biggest killers in industrialised countries such as the US. The rates of obesity are highest among those on the lowest income, who subsist in 'food deserts' on low-value diets.

That is, they have limited access to shops selling healthy foods such as fresh fruit and vegetables and grains. Although there is more cheap food, it is more processed and less healthy.[22] New Internationalist said in 2011 that, of a world population of seven billion, one billion ate too much and one billion lived in food poverty.[23] 'The EAT-Lancet Report' suggests that a shift to a healthier, more plant-based diet will be not only better for our health but also a necessary part of food production systems that can be sustained in the long term. This diet would consist of 'half a plate of fruits, vegetables and nuts. The other half consists of primarily whole grains, plant proteins (beans, lentils, pulses), unsaturated plant oils, modest amounts of meat and dairy, and some added sugars and starchy vegetables.'[24] Eating meat is still a healthy option here but the emphasis is on less of it and on poultry rather than red meat and pasture-fed livestock rather than grain-fed livestock.

Meeting the four challenges: the characteristics of sustainable farms

To meet these four challenges, farms and food systems will need to reduce those practices which are most harmful, and replace them with regenerative farming methods that serve the same function. What might this look like? The story of this book is that these farms already exist, but on a tiny scale, in the form of biodynamic, organic, permaculture, agroforestry, agroecology and regenerative farming systems. These systems have been pioneered, researched and proven but only on a very small scale and are often dismissed because they have arisen from a different paradigm of thought. ■

1 Attributed to Goethe.

2 DEFRA, 'Agricultural Statistics and Climate Change', 2020.

3 https://www.farmcarbontoolkit.org.uk

4 T.F. Stocker, D. Qin, G.K. Plattner, M. Tignor, S.K. Allen, J. Boschung, A. Nauels, Y. Xia, V. Bex and P.M. Midgley (eds), *Climate Change 2013: The Physical Science Basis,* Cambridge University Press, Cambridge, 2013.

5 L. Wellesley, C. Happer and A. Froggatt, *Changing Climate, Changing Diets,* Chatham House, London, 2015.

6 J. Pretty, *Agri-Culture,* Earthscan, London, 2002.

7 FAO, 'Climate Change, Agriculture and Food Security', 2016.

8 DEFRA, 'Agricultural Statistics and Climate Change'.

9 P. Roberts, *The End of Oil,* Bloomsbury, London, 2004.

10 UN Climate Change, 'Paris Agreement on Carbon Reductions'.

11 FAO, 'Climate Change, Agriculture and Food Security'.

12 Pretty, *Agri-Culture*; UN Climate Change, 'Paris Agreement on Carbon Reductions'

13 Pretty, *Agri-Culture.*

14 RSPB, 'State of Nature Report', 2019; FAO, 'Status of the World's Soil Resources', 2015.

15 FAO, 'Climate Change, Agriculture and Food Security'; P. Holden, 'Landsparing vs Landsharing', Sustainable Food Trust, 2020.

16 Ocean Cleanup, 'Marine Plastic Pollution', 2020.

17 S.H. Swann and S. Colino, *Countdown: How our Modern World is Threatening Sperm Counts,* Simon & Schuster, New York, 2021.

18 Pretty, *Agri-Culture*; RSPB, 'State of Nature Report'.

19 FAO, 'Climate Change, Agriculture and Food Security'; FAO, 'The State of Food and Agriculture: Moving Forward on Food Loss and Waste Reduction', 2019.

20 FAO, 'The State of Food and Agriculture'.

21 E. Cassidy, P. West, J. Gerner, J. Foley; 'Redefining Agricultural Yields: From Tonnes to People Nourished per Hectare', *Environ*mental Research Letters. 8, 2013, pp. 1–8, doi:10.1088/1748-9326/8/3/034015 Institute on the Environment (IonE), University of Minnesota, Saint Paul, MN 55108, USA.

22 B. Popkin, L. Adair and S. Ng, 'Now and Then: The Global Nutrition Transition: The Pandemic of Obesity in Developing Countries', *Nutrition Review* 70, 2014, pp. 3–21.

23 V. Baird, *No Nonsense Guide to World Population*, New Internationalist, Oxford, 2011.

24 EAT-Lancet Commission, 'The EAT-Lancet Report', 2019.

Chapter 2

How did we get here?

How did the industrialised food production system come about?

To understand more fully the challenges that food production faces today, we need to understand how we got where we are. The Talking Heads line 'And you may ask yourself, well, how did I get here?' was one of my favourite lyrics when I was at university studying for a degree in horticulture. Even in the 1980s, what I was taught did not seem to make much sense in the face of the onset of environmental crises. We find ourselves in the wrong place with food production because of a way of thinking that does not consider the complexity of the food system as a whole. The focus has only been on yields and profit as outcomes, and not on the multiple yields that farming and food production provide. To put it another way, food production has not been thought of as a 'holistic system'. Planetary and human health have not been included in the equation. We are trying to solve the problem with the same set of principles that created it, and this will not work. Modern sustainable food production systems provide multiple yields: they produce food, they produce and support biodiversity, they sequester carbon and they don't pollute watercourses with nitrate runoff. The cost of cleaning up after industrial farming is estimated to be £209 per hectare per year.[1] Industrial food production systems externalise their costs: they don't pay for watercourse pollution from nitrate fertiliser runoff; they don't pay for damaging people's health with carcinogens; they don't pay for the loss of biodiversity. Not only do they not pay for their pollution, but in most countries they are subsidised to do the polluting, and so the taxpayer pays twice over. The 'polluter pays' economic principle is not applied here in any shape or form.

Full understanding of the development, drivers and consequences of the industrial food system reveals that this system is simply not fit for purpose. A complete systems rethink is required. Instead of making a few tweaks to the current system, a whole new paradigm of thought is required. This paradigm started to develop in response to industrial farming and the introduction of nitrate fertilisers. The ecologically integrated paradigm has continued to develop ever since, but on the margins. It's upon the principles and practices of this paradigm, that has been developed over the last 100 years in this divergent tradition of sustainable food systems, that we can build truly modern healthy, regenerative and sustainable food systems.

The agricultural revolutions

The first agricultural revolution started in England, arguably from the 14th century onwards. It paved the way for the industrial revolution in the 1850s by freeing up thousands of people from farming to migrate to the cities to work in the factories and mills. The yeoman farmers left behind in the countryside established small mixed farms, producing enough cereals, flour, beef, sheep, milk, eggs, vegetables and fruit to feed the local rural and urban populations.

There was a move from a community-based, albeit feudal system of farming to the 'enclosure' of common land into individual private ownership and use. Over hundreds of years, the land was privatised to create in England the hedged fields systems that we now recognise. Such 'enclosures', as they were called in England, were more accurately called 'clearances' in Scotland, where peasant farmers were forcibly cleared from the land. Crops became increasingly commodified as the farms became bigger and landless peasants were shipped off to colonise other parts of the world or went to work in urban factories.

The enclosure process has been repeated across the world to this day. Peasant subsistence farmers are moved off the land that they have farmed and lived on for generations, to make way for large-scale industrial farming and agribusiness for profit. Colonising Europeans replicated European models of farming in North America, South America, Africa, India, Indonesia, Australia and New Zealand, causing

huge damage to ecosystems and pushing indigenous peoples from their lands and largely ignoring their age-old systems of sustainable food production well suited to their soils and climate.

Slavery and the colonisation in the Americas were the very first step in the industrialisation of food, specifically sugar and tobacco as export cash crops. Based on the exploitation of millions of Africans, rather than on fossil fuels, this system generated the wealth that financed the industrial revolution.

The second agricultural revolution started in Europe in 1913, a time of high industrial unemployment coupled with a farming depression and low food prices caused by the import of cheap food from the new European colonies around the globe. Low food and land prices in Europe meant that, to survive economically, farmers had to intensify their methods of food production. The old ways were dying out, as beautifully recorded by Ronald Blythe in Akenfield.[2] New methods of industrial farming were pioneered using machinery, the provision of fertility out of a packet, and new ways of combating pests, diseases and weeds. Fertilisers and pesticides were developed out of the same technology that produced weapons used in World Wars I and II.

The Haber process was developed in 1909 and enabled nitrate fertilisers to be manufactured from 1913 onwards to increase productivity in farming. This was a key factor in the second industrial agricultural revolution and has to this day facilitated the increasing agricultural yields needed to support the world's growing population. The Haber process converts gaseous nitrogen from the atmosphere,

The Soil Food Web

Nematodes
Root-feeders

Arthropods
Shredders

Arthropods
Predators

Birds

Fungi
Mycorrhizal fungi
Saprophytic fungi

Nematodes
Fungal- and
bacterial-feeders

Nematodes
Predators

Plants
Shoots and
roots

Organic
Matter
Waste, residue and
metabolites from
plants, animals and
microbes.

Bacteria

Protozoa
Amoebae, flagellates,
and ciliates

Animals

First trophic level:	Second trophic level:	Third trophic level:	Fourth trophic level:	Fifth and higher trophic levels:
Photosynthesizers	Decomposers Mutualists Pathogens, Parasites Root-feeders	Shredders Predators Grazers	Higher level predators	Higher level predators

United States Department of Agriculture

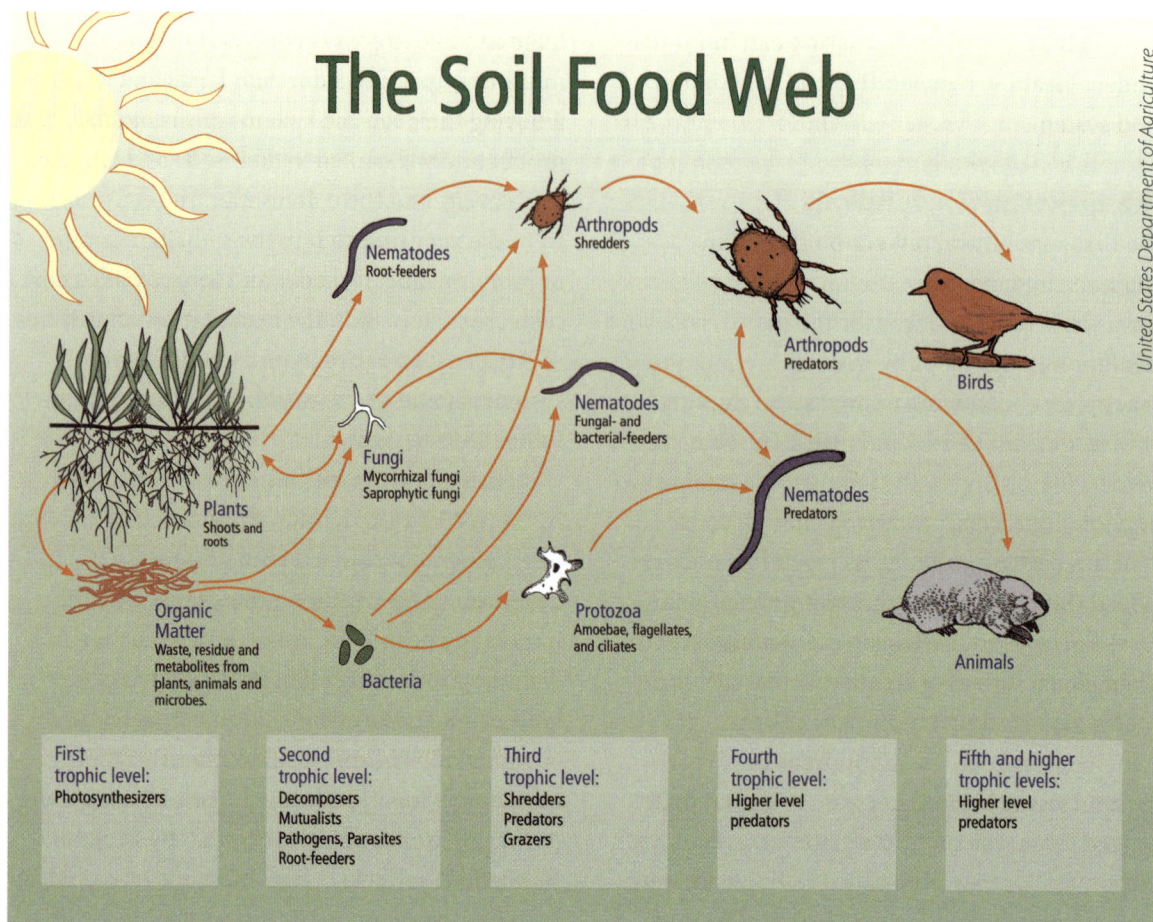

Figure 2.1: The soil food web.

using fossil fuels, into nitrates. These are solid soluble forms of nitrogen that are used as fertilisers. Nitrous oxide is emitted in the process. Nitrogen is a macronutrient for plants and it fuels plant growth. It is a building block of protein formation in plants using the energy from photosynthesis. Before the Haber process was developed, nitrates were only available from manure and legume crops and from naturally occurring bacteria in the soil. They were regarded as the limiting factor in agriculture at that point, since they are highly soluble and will leech out of the soil. The other macronutrients – phosphate and potassium – are insoluble, occur naturally in the soil and are generally not in short supply. The development of a synthetic nitrate fertiliser allowed the decoupling of livestock from farms, since their manure was no longer needed. The production of nitrate fertilisers contributes significantly to climate change, by emitting six per cent of global GHG.[3] Runoff of nitrates into rivers and streams is a major pollutant, killing fish. High nitrate levels in food are a proven carcinogen.[4]

When artificial nitrate fertilisers are applied to the soil they kill off many of the microorganisms found there, especially the nitrifying bacteria. These form part of the soil's complex ecology. Once the food chain is broken, the natural ecology of the soil starts to unravel and the microbial activity in the soil is reduced. It was only in 1996 that Elaine Ingham, with her 'soil food web' research (Figure 2.1), discovered exactly how the soil ecology works.[5]

Nitrates from the Haber process were also used to develop bombs. Haber went on to develop organochlorines used as nerve gases in World War I and in the gas chambers in World War II. Organochlorines and organophospates form the basis for many insecticides. One of these, DDT, has accumulated in soils and food chains across the globe.

The development of refrigeration, lorries, trains, ships and planes meant that food could be transported large distances across the world, and stored for long periods of time, thereby decoupling food from season and place – all by the power of fossil fuels. This food needs to be wrapped and packaged in plastic to help it withstand the journeys. We can now expect to walk into a supermarket and obtain avocados or New Zealand lamb – or buy green French beans flown in from Kenya all year round.

World War II brought widespread hunger, malnutrition and starvation across continental Europe. The end of the war saw the formation of the precursor of what is now the European Union (EU), to prevent any further outbreak of war in Europe. One of the founding principles of the EU was the Common Agricultural Policy (CAP). This stipulated that there should always

be enough food for all. Large subsidies were paid to farmers to implement industrial farming methods: ripping out hedges, draining marshes and meadows, cutting down ancient woodland and ploughing heathlands. Farmers were encouraged to apply artificial nitrogen fertilisers and to use pesticides. Livestock were put into large sheds and industrialised, as for example in battery farming of chicken and in feedlots where cattle are fed with grain and other waste products and artificially inseminated. At this time the problems associated with these methods were unknown.

The psychology of scarcity helps us to understand the rollout of industrial farming. This can be illustrated by what happened in the Netherlands in 1944. The Dutch population collectively experienced the Winter of Hunger – the last winter of the war, when the entire population was put on a ration of 500 kilocalories per day. The lack of food resulted from the disruption of farming by the war. Imagine living through one of the coldest winters of the century on 500 kilocalories per day, being cold and always hungry, with no electricity most of the time and having to cycle a bike with a dynamo indoors in the evening to be able to read. This time is often referred to as a 'famine' and many Dutch people suffered from malnutrition.

After the war, the CAP subsidised the implementation of industrial farming systems and dramatically increased food yields. The US rolled out industrial farming around the same time, enabling massive subsidised overproduction of food for many years. In Europe excess production was known as the 'butter mountain' and the 'wine lake'. This line of action was an understandable response to the experience of large-scale food shortages. Data collected from the Dutch population provides the basis of long-term studies of the effects of the severe but short-term food shortage during the war. It was found to have triggered long-term anxiety, depression and eating disorders.[6]

By the 1990s it was common to feed cattle ground-up cattle proteins, including brain tissue. This caused 'mad cow disease' (or bovine spongiform encephalopathy – BSE), which transferred to people who ate this infected meat, causing Creutzfeldt–Jakob disease (CJD). New crop varieties and animal breeds were bred as high-yielding varieties (HYVs) for industrial systems that require high levels of fertiliser and pesticide. Yields were rapidly increased by 50–100 per cent, but these yields have begun to decline as soils have become depleted and the pests and

Ecopsychology perspectives on industrial food systems

Mary Jayne Rust and other ecopsychologists explore the mindset that keeps us trapped in unsustainable practices – how, collectively or individually, we are unable to change our behaviour towards the environment even when we know this behaviour is unhealthy for us. They also develop methods of working with nature to restore mental health.

When people have suffered the trauma of a shortage of food, they compensate with an unhealthy relationship to food. Eating disorders become common, as do anxiety and depression, as we can see from the longitudinal studies in the Netherlands. The memory of the trauma can be passed down the generations, by culture, but also by our genes. The study of epigenetics has found that trauma switches on genes that pass down traits to the next generation.

One key issue of the health epidemic in the global north is the overeating of overprocessed food, which leads to obesity, heart disease and cancers. Might this be seen as a collective unhealthy relationship to food? Ecopsychologists conjecture that people in the global north may have collectively suffered trauma that has impacted on their relationship with food.[7]

Ecopsychologists conjecture that it may be the trauma of past food shortages that keeps us so addicted to industrial food systems, that is, to having lots of cheap food available all the time, and to inhabiting the psychology of scarcity rather than the psychology of abundance.

The last few hundred years have seen frequent collective traumas involving food, including the loss of land and the means to feed oneself. During these centuries, slavery and racism have traumatised millions of people of African descent. In the 19th century the potato famine in Ireland and land clearances of Scotland triggered mass emigration to the US, Canada, Australia and New Zealand. In the early 20th century, shortages of food in Italy drove mass emigration to the US. The 1930s saw the Great Depression, poverty and food shortages in the US and in Europe. The great Dust Bowl disaster in the US weeds have become immune to the pesticides. The foods produced have also shown a drop in their mineral content.[9] However, not only was there more than enough food for the world's population, but the food became cheaper, the average household spending less on their weekly food basket than ever before. At the same time, the percentage the farmer receives from the sale of the food has decreased, from 45–60 per cent 50 years ago to seven per cent today.[10]

Across the world, industrial food systems have been rolled out and labelled the 'green revolution'. We could call this the third agricultural revolution. Indigenous farmers in the global south have been introduced to nitrate fertilisers, pesticides, new seed varieties (HYVs) and genetically modified organisms (GMOs), all of which increase yields for a few years.

The new seeds are expensive and patented to a few giant agribusiness companies. The 'nutrition

exacerbated food shortages. World War II saw mass emigration to the US of Jews and others fleeing a devastated Europe, where hunger and malnutrition were rampant.

Ecopsycholgists suggest that collectively we have become so fearful of food scarcity that we have become addicted to the methods of production that enable us to eat a fast-food burger any time of the day or night for a tiny amount of money.

Rust suggests that in order to heal an unhealthy relationship with food an individual must first process the trauma they have experienced and reconnect the sensations in their body, such as hunger, or feeling full after a meal, to their eating habits. That means developing an awareness of when they are hungry and when they are full. What might this kind of process look like on a collective level? Maybe it would involve reconnecting to the source of our food by means of a seasonal local diet. Maybe baking and cooking from scratch. Maybe celebrating food with a special meal or event with our family and friends. Maybe farmers will have to change their approach to farming. Maybe everyone will have to change their mindset from that of the psychology of scarcity – the fear there won't be enough food – to one of the psychology of abundance, that there can be enough for everyone, and for biodiversity, if we take only what we need and if we reduce waste. This is the mindset celebrated by La Via Campesina in their Food Sovereignty declaration in 2007:

> *Our heritage as food producers is critical to the future of humanity. This is especially so in the case of women and indigenous peoples who are historical creators of knowledge about food and agriculture and are devalued. But this heritage and our capacities to produce healthy, good and abundant food are being threatened and undermined by neo-liberalism and global capitalism.[8]*

transition' that has followed the industrialisation of agriculture has increased demand for meat in many countries. In some cases, the industrialisation of beef cattle has created giant feedlots where cattle are fed on grain imports from across the globe, the resulting beef is exported across the globe and they produce methane that contributes to climate change.

The process of industrialisation is continuing today. The so-called 'fourth agricultural revolution' includes the introduction of more GMO plants and livestock and the genetic editing of existing crop varieties and animal breeds. These are high-tech solutions to deal with the problems arising from the industrialised food production. Excessive use of nitrate fertilisers in arid climates causes the salinisation of soils. Modern crops have genes spliced into them to enable them to grow in these conditions. Other crops are spliced with genes resistant to the glyphosate herbicide, which means the farmer can spray the entire crop to kill off the weeds but not the crop itself. However, this practice also allows the herbicide to enter the food chain. Meanwhile the technology is being developed to replace meat from animals with cultured meat grown in laboratory conditions from muscle stem cells. Vertically farmed vegetables, grown in hydroponic systems on stacked-up shelves under artificial light, will substitute plants grown in fields with sunlight. Robots weeding and picking produce are set to replace people in the fields.[11]

It is between 1850 and the present day that we see the sharp rise in the IPCC graph of carbon emissions, corresponding to the world's ongoing industrialisation. Rachel Carson was one of the first, in 1960, to notice the 'silent spring' as bird populations were decimated by lack of food and habitat and by devastating pesticides.[12] The observations of biodynamic farmer Marjorie Spock on Long Island, New York, helped to inform Carson's seminal book, *Silent Spring*. The steady march of the loss of biodiversity has carried on since then.

Changes in the food production systems are now urgent. We could call the changes now required the 'sustainable agricultural revolution' (or the 'agrarian renaissance', as Colin Tudge calls it). This transition would produce enough healthy food for everyone to eat, while supporting biodiversity, sequestering carbon and reducing fossil fuel use, and can do all of this in an era of erratic weather and climate change. Its beginnings are all around us, on the edges of farming communities and in indigenous systems of farming.

Productionist and ecologically integrated food systems

There have been two parallel systems of food production over the last 100 years. With every step down the route of industrial food production there has been an alternative step down the route of sustainable food production. Sociologists refer to them as, respectively, the 'industrial productionist' paradigm and the 'ecologically integrated' paradigm. Martin Wolfe, plant pathologist and agroforestry pioneer, suggested that there are two ways to approach the creation of sustainable farms. One is to bolt on solutions to whatever problems are encountered in continuing to pursue the

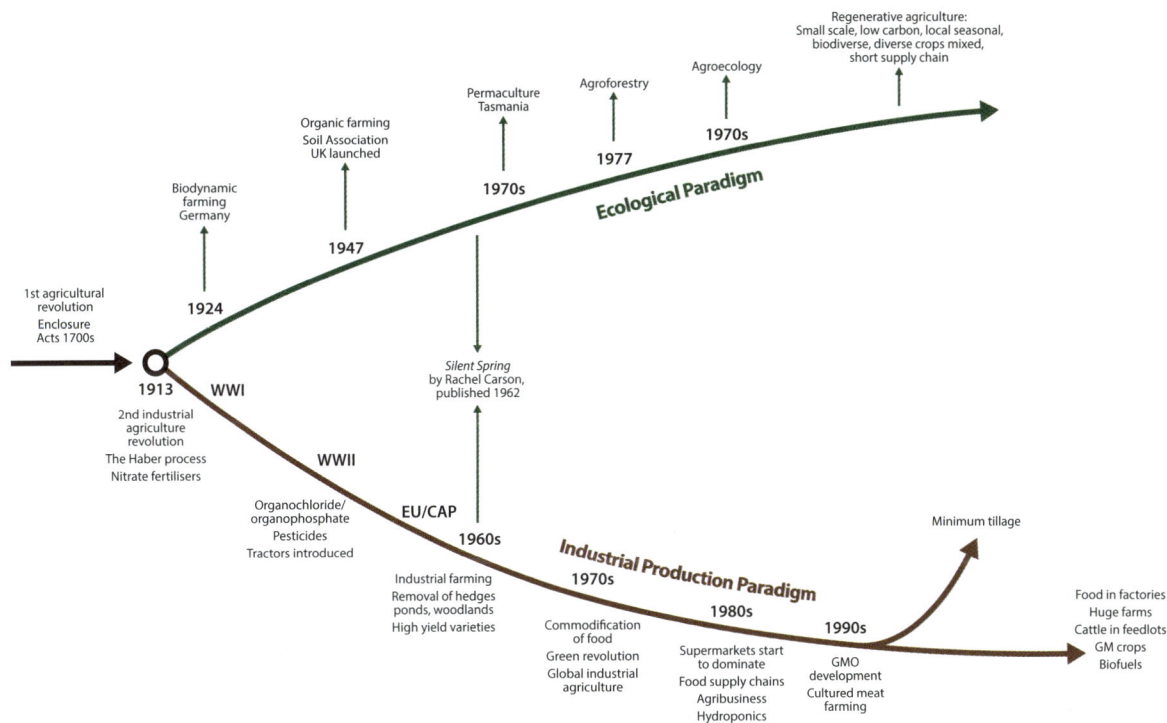

Figure 2.2: Conventional versus sustainable food production methods.

dominant industrial paradigm. The other is to look back to the point of divergence and explore the pathways of the different models of sustainable food production. The sustainable food systems draw upon millennia of practice in indigenous food systems as well as new modern developments. They can be combined as a toolkit to build the truly modern farms for the future which will meet all of the four challenges listed in Chapter 1. All the farming systems covered in this book have arisen from a complete systems rethink. They started to emerge at the point of divergence from the industrial tradition, in 1913 (Figure 2.2).

If we bolt regenerative practices on to the industrial systems that are currently in place then one or two of the challenges can perhaps be addressed but other problems will pop up.

It's a bit like squeezing a balloon: when one part is squeezed, the air is displaced to form a bulge in another part. Industrial farming systems aim to increase food production and yields in order to feed more people by bringing more land into production, increasing the use of nitrogen-based fertilisers and the introduction of GMOs. Such measures will bring yet further collapse in biodiversity and the environmental services upon which everyone depends. We can produce more biofuels to reduce the demand for fossil fuels in transport, but then there will not be enough land left to produce enough food. We can grow meat in factory conditions but this is energy intensive and produces toxic slurry. The industrial solutions to the problems just create new problems.

In the sustainable food production paradigm the problems are tackled head-on. The methods used gradually regenerate soils, produce resilient food systems that create resilient economies and healthy people. They rebuild biodiversity and watersheds. They are also very good at sequestering carbon.

What is a sustainable or regenerative food production system?

To create sustainable food systems we need first to build a picture of the positive things farming can do to promote sustainability. We need to reduce those farming activities which do not promote sustainability. We need to think of sustainability as a process rather than a state. Farms can become 'more' sustainable and then keep taking steps in the same direction. It can be a gradual change, or it can be a fast change. For example, a heated glasshouse producer growing early season strawberries for a supermarket in peat substrates can take the first step in reducing fossil fuel use by switching the heating system from oil to biomass. In 10 years' time they may be growing 10 different crops, not just strawberries, and these may be seasonal crops that don't need heating and are grown in the soil rather than in peat- or coir-based substrates. This kind of trajectory moves farms in the direction of becoming more sustainable and less harmful. Really damaged soils and ecosystems will need some form of regenerative agriculture to repair the damage.

The main characteristics of all the sustainable farming systems that need to be developed are that they:

- are closed loop and return fertility to the soil
- build organic matter in the soil
- resilient to climate change because they have organically rich soils
- collect rainwater
- are home to lots of trees
- are polycultural rather than monocultural
- produce abundant delicious seasonal nutrient-dense food
- sell into short supply chains
- are plastic-free
- are low on waste
- are full of biodiversity
- are economically resilient
- employ many people
- are great places to work
- often, but not always, include livestock.

These positive characteristics are found in abundance in biodynamic, organic, permaculture, agroforestry, agroecological and regenerative farming systems. These types of systems currently account for a very small percentage of the world's agriculture (Figure 2.3).

The things that farms for the future will need to stop doing are:

- using nitrate fertilisers and any pesticides
- airfreighting unseasonal produce around the world
- heating glasshouses with fossil fuels
- monocropping
- clearing forests, woodlands, ponds and wetlands
- using plastic packaging
- wasting food
- rearing livestock and poultry in intensive systems

Current state of sustainable farming

6% Agroecology
1.5% Organic and biodynamic
92.5% Industrial or unimproved

What if we push the slider up a bit?

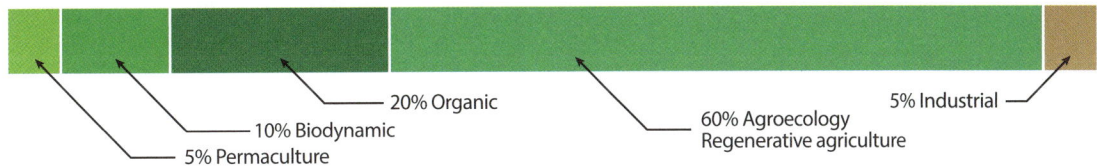

20% Organic
10% Biodynamic
5% Permaculture
60% Agroecology
Regenerative agriculture
5% Industrial

Source: www.statistics.fibl.org.

Figures 2.3 and 2.4: The current scale of sustainable farming in the UK, and Europe and the rest of world.

- shipping animal feed around the world
- using GMOs
- exploiting agricultural workers
- paying low wages.

Rob Hopkins, one of the founders of the Transition Towns movement, talks about giving ourselves time and space to ask the question 'What if?' This 'helps us unlock the imagination in service to the big challenges we face'. What if half or even all of our food came from sustainable sources? What if we went back to the point of divergence and redesigned all our food systems to be truly sustainable? What would that look like? How would it affect our health? How would the food taste? How many birds would there be? What would shopping for food be like? What would it be like to be a farmer or grower? Where would the food be grown? In cities as well as rural areas? Imagine a world where we move the 'slider' up on Figure 2.3 and ask what the world would look like if

50 per cent, or more, of all food systems were sustainable (see Figure 2.4).[13] This may seem like an impossible dream, but even the FAO suggests that all farming in the world needs to transform into sustainable agroecological farming in order both to mitigate and to adapt to climate change:

A broad-based transformation of food and agriculture systems to sustainable agriculture systems is needed to ensure global food security, provide economic and social opportunities for all, protect the ecosystem services on which agriculture depends, and build resilience to climate change. Without adaptation to climate change, it will not be possible to achieve food security for all and eradicate hunger, malnutrition and poverty.[14]

There are two countries in the world that have done this: Cuba and Switzerland. Cuba had to decarbonise its farming systems when it lost its supply of fossil fuels in the 1990s because of the collapse of the USSR. It accomplished this in three years. Switzerland chose a route

Julie Brown's vision for food production in the UK in 2035

It's 2035 and a network of farms in and around urban areas provides at least 60 per cent of the food needs of those towns and cities: 20 per cent comes from other parts of the UK and a further 20 per cent comes from overseas.

At least 20 per cent of people in the UK are now involved in farming in and around the area where they live, and most people's portfolio of work includes some food-related work – paid and unpaid – that helps enable them to live. Farming continues to be about producing food for a living, but this increasingly includes such varied approaches as part-time patchwork urban farming and the involvement of city people in running orchards and farms that supply food to their communities.

All food producers, wherever they live, are paid a fair price for their produce.

Many people are also involved in growing some of their own food, but more significant is that they are connected to the people and the farms (urban or rural) that produce most of the food they eat. People regularly volunteer at their local urban market gardens, or contribute some or all of their garden to a patchwork farm, or grow a proportion of their own food on an allotment. They understand where the rest of their food comes from and how and by whom it is produced. They value and respect its producers. After all, they know what skill it takes to grow food all year round, particularly now the weather is so unpredictable. They buy fresh food from community-led box schemes, markets, community shops, CSA schemes[15] and online schemes and they know how to cook and enjoy that food. Fish schemes like Soleshare abound,[16] as do artisan bakeries like E5 Bakehouse and Better Health as well as local distillers and brewers.[17]

Supermarkets still exist but they control a much smaller share of the food market and specialise in consumables and processed foods, leaving most of the fresh produce to traders better able to distribute it in a way that meets the needs of the farmers. The food the supermarkets sell comes from the remaining larger, more industrialised farms, for whom this method of distribution still makes some sense, and they are able to provide some of the more basic products – such as bread, butter and cheese. Many supermarkets have returned to being what they were originally – a one-stop shop under one roof – but one where produce is sourced along the same lines as in the community-led systems.

Organic certification is no longer necessary, since all food production is pretty much organic. Emergency use of certain pesticides and temporary use of artificial fertiliser do happen but as the exception rather than the rule.

Farmers sell through a combination of routes. Supply chain cooperation – through regional networks of traders – means the traders' journeys are minimised and loads are optimised. The network of Better

Food Traders grew out of the Growing Communities Start-Up Programme in 2016.[18] Many thousands of other box schemes, markets, shops and online retailers have since joined the network.[19] Regional depots or hubs, such as the Better Food Shed, provide central drop-off points for producers, along with other services.

Farmers also cooperate with each other more than they used to – sharing knowledge as well as larger bits of equipment and processing facilities. The Landworkers' Alliance has grown in size and now represents and campaigns on behalf of thousands of farmers and growers.[20]

The populations of urban areas have reduced somewhat as people have moved to the countryside to produce food. This has reinvigorated the rural towns and villages as well as freeing up urban land on which to grow food, so that urban areas are able to make a bigger contribution to feeding their own populations. The reduced need for car parks and roads has freed up even more land for urban food production.

People eat out a lot – communal one-pot affairs, since it makes sense to pool resources and it's nice to eat with your community. Not all the time, though! Sometimes people splash out on more bespoke places where they can get away from their neighbours! There is far less takeaway and convenience food – it's just too energy intensive – and people are happier to eat at home, since they know how to cook and are less rushed for time now that working patterns have shifted.

People eat a lot less meat. Any protein gap is filled by beans and pulses – from traders like Hodmedods[21] – and from urban aquaponic and insect-based systems. Urban chickens and goats abound – although they are kept mainly for their eggs and milk and rarely end up in the pot until they are quite old.

Everyone knows their own fair allocation of food in terms of calories, optimum nutrition and environmental impact. A clever app and coding system enables people to shop as they please while keeping a tally of their budget. People can trade on any spare allocation or save up for a blowout. A similar system for transport was introduced in 2025 so that if you cycle, walk or take public transport most of the time you can save up for, say, a flight at some date. You can also trade across sectors, which means that vegans get to take many of the flights available!

In Growing Communities' veg scheme an app is used to check produce before it is delivered for packing, so nothing is ever sent back or requires extra sorting time. Nifty sensors in the packing area count the bag numbers and ensure that numbers on the delivery vehicles and at the collection points tally with the packing lists.[22]

And what about me? As a sprightly 72-year-old, I have finally learned how to grow pears and in my spare time I've taken to making knitted vegetables and little origami animals.[23]

of food security by means of a farming policy supporting a high level of sustainable farming. Both these examples are discussed in greater depth in Chapter 11.

Critics frequently ask how the sustainable methods of food production will feed the world. It is possible to produce enough food, as will be explored in Chapter 9 and illustrated using a case study of Huxhams Cross Farm in Chapter 10. Chapters 3–8 will describe sustainable farming systems that have existed since 1924. Scaled up, these systems are the basis of the toolkit that could transform farming enterprises into sustainable systems over the course of the 21st century.

In the meantime, Julie Brown, the Director of Growing Communities in London, is pioneering and recreating local food supply chains into London. Above is her vision for food production in a large city in 2035. ■

1   J. Pretty, *Agri-Culture*, Earthscan, London, 2002.

2   R. Blythe, *Akenfield*, Penguin, London, 1969.

3   T.F. Stocker, D. Qin, G.K. Plattner, M. Tignor, S.K. Allen, J. Boschung, A. Nauels, Y. Xia, V. Bex and P.M. Midgley (eds), *Climate Change 2013: The Physical Science Basis*, Cambridge University Press, Cambridge, 2013.

4   RSPB, 'State of Nature Report', 2019

5   https://www.soilfoodweb.com

6   https://www.hongerwinter.nl/

7   M.J. Rust, 'Nature Hunger: Eating Problems and Consuming the Earth', *Counselling Psychology Review* 23(2), 2008, pp. 70–78; M.J. Rust, 'Climate on the Couch: Unconscious Processes in Relation to Our Environmental Crisis', *Psychotherapy and Politics International* 6(3), 2008, pp. 157–170.

8   https://viacampesina.org/en/food-sovereignty

9   Rust, 'Climate on the Couch'.

10 T. Lang and M. Heasman, *Food Wars*, Earthscan, London, 2004.

11 M. Gove, 'Welcome to the Fourth Agricultural Revolution', speech at Oxford Farming Conference, 2019.

12 R. Carson, *Silent Spring*, Penguin, London, 1962.

13 R. Hopkins, *From What Is to What If*, Chelsea Green, Hartford, VT, 2019.

14 FAO Report, 'Climate Change, Agriculture and Food Security', 2016

15 https://communitysupportedagriculture.org.uk

16 https://www.soleshare.net/#home

17 https://e5bakehouse.com/; https://www.betterhealthbakery.org.uk

18 https://growingcommunities.org/

19 https://laruchequiditoui.fr/fr; https://www.openfoodnetwork.org.uk

20 https://landworkersalliance.org.uk/

21 https://hodmedods.co.uk/

22 https://growingcommunities.org/organic-veg-scheme/

23 https://growingcommunities.org/

Part 2

systems of
sustainable
and
regenerative
food production

Chapter 3

Biodynamic food production

In modern times, people have lost all insight into the great interrelationships of nature… into how it is common to plants and to the soil and to the excretory products of life as we find them in manure.[1]

Rudolf Steiner

Introduction

The first of the sustainable farming systems we shall examine was born in direct response to the introduction of industrial farming methods in Germany. Biodynamic farming came into being from a series of eight lectures given by Rudolf Steiner at Koberwitz in 1924. He had been invited to a large estate, then in Germany but now in Poland, at the request of a community of farmers after they noticed a decline in the fertility of their soils. The lectures were attended by about 100 people.[2]

This chapter will explain the principles and practices of biodynamic farming, what long-term effect they have on soils and the farming community, what kind of food they yield and where you can find biodynamic farms. The ways that biodynamic farms meet the challenges of mitigating and adapting to climate change, offsetting biodiversity and

producing food will be explored further in Part 3. Finally, there are pointers to where you can find out more about biodynamic food production systems. Biodynamic farming will be illustrated by a case study of two successful biodynamic farms. A third case study, Huxhams Cross Farm, is explored in depth in Chapter 10.

What is biodynamic farming?

Since the original lectures by Steiner, biodynamic farming and gardening have developed into a clear methodology that is described in the Biodynamic Federation–Demeter International's Standards of Farming. These are accredited by the International and European Regulations for Organic and Biodynamic Methods of Sustainable Farming and regulated by regional Biodynamic Associations found across the globe, though mostly in Europe, the US, Australia and New Zealand. Biodynamic food is sold under the trademark of the Demeter logo, which was the first recognised symbol for any ecological product (Figure 3.1).

Demeter is the Greek goddess of harvest and fertility. The headquarters of the biodynamic movement is based in the

Goetheanum in Dornach, Switzerland.

Biodynamic farming is practised in 65 countries. There are over 6,500 Demeter farms and nearly 221,000 hectares of cropping in total. There are currently 100 registered biodynamic farms in the UK, covering 3,886 hectares. Examples of good practice include the 200-hectare community-owned Tablehurst Farm in Sussex. Germany has the most biodynamic farms: 1,700 of them, covering 93,000 hectares. The US has just over 150 biodynamic farms covering 5,000 hectares in total. Sekem in Egypt is an exemplar farm that started on 70 hectares and now cooperatively farms more than 2,000 hectares; it produces cotton, herbs and food and runs an educational programme. In Sri Lanka biodynamic farming is practised on over 1,400 farms. In China the BD movement is embryonic; just a few farms are operational.[3]

Where did biodynamic farming start?

Why did that group of farmers in Germany ask Steiner for guidance on their soil fertility and why was the soil fertility in decline in Central Europe by 1924? As we saw in Chapter 2, after the development of the Haber process in 1909, nitrate fertilisers began to be manufactured in Germany and used on German farms, and these fertilisers had a destructive impact on the microbial activity in the soil. Steiner said in his lectures, 'no one realises today that all the mineral fertilisers are just what are contributing most to the degeneration of the products of agriculture'.[4]

The farmers in Koberwitz had both the observation skills and the foresight to seek

Figure 3.1: Biodynamic food is sold under the trademark of the Demeter logo.

guidance about the problems that were starting to arise from the early use of nitrate fertilisers. They spoke of being able to grow the same crop on the same piece of land for up to ten years, this is unthinkable practice in the current rotational systems. No doubt many farmers noticed the vitality of their crops dropping, but the magic of putting on fertiliser with machinery and getting high yields must have seemed like a great boon after the slog of manual farming.

Instead of going back to previous ways of farming, the farmers sought an alternative. This led to today's biodynamic farming and also the basis of organic farming.

Steiner established the broad outlines of biodynamic farming in his eight lectures. At his request, however, Ehrenfried Pfeiffer had started experimenting with practices of this kind two years previously. Pfeiffer was the head of the Natural Sciences section of the Goetheanum in Dornach. After Steiner's lectures, Pfeiffer went on to experiment further on a 200-hectare farm at Loverendale in the Netherlands. He regularly met with a group at Dornach called the Experimental Circle and they produced the first set of standards in 1928.

Pfeiffer was invited to the UK in 1939 by

Lord Northbourne, who was hosting a nine-day conference on biodynamic farming on his estate at Betteshanger in Kent. Dr Scott Williamson of the Peckham Experiment (see Chapter 11) was also at this conference. After this event Northbourne wrote *Look to the Land* in 1940, in which he coined the phrase 'organic farming'. Eve Balfour, who also attended the conference, subsequently published *The Living Soil* in 1943. It was from this meeting and these books that the organic movement was born in the UK.[5] Hence the connection between the two systems. John Paull suggests that Northbourne deliberately removed anthroposophy from the biodynamic methods to formulate a more secular system, which he named 'organic'.

The conference was held just a few months before the outbreak of World War II.[6] Pfeifer and his family moved to the US in 1940. Pfeifer started a farm in Chester, New York, and founded the US Biodynamic Association. He also worked with Jerome Rodale, who founded the organic movement in the US. Once again we see how the biodynamic movement was the precursor of the organic movement.[7]

Why did the German farmers ask Steiner, who was not a farmer, to devise a new farming system? Steiner was born in 1861 in Croatia, then in the Austrian Empire. He died in Switzerland in 1925, soon after giving the biodynamic lectures. In total he wrote 30 books, gave 6,000 lectures and developed the philosophy called 'anthroposophy'. This translates literally as 'the wisdom of the human being' and can be described as 'The path of inner development that aims to guide the spiritual in the human being to the spiritual in the universe'. Steiner studied Goethe's scientific methods and was a holistic scientific researcher with keen observational skills.

He founded Waldorf education, now a global educational movement, and inspired a special needs education and therapeutic movement, called Camphill, and holistic medicine, dance and drama movements. The son of a railway worker, he grew up in a rural area and records that he was well acquainted with the rural traditions of the time: 'All these things [agriculture] were part of my life for a long time… the fact that I grew up surrounded by agriculture means I grew up loving it.'[8]

Steiner described himself as a scientist and a clairvoyant. He used a method of spiritual contemplation to gain insights that he brought back from 'super-sensible worlds'. The word 'seer' or 'shaman' might be used to describe someone able to travel to a spirit world and bring back information to guide or heal a community. Arnold Mindell, a Jungian psychotherapist, describes the shaman as someone who 'always takes some form of psychic journey to the world of spirits to find what is missing in everyday life'.[9]

To fully understand biodynamic agriculture requires most people to look at the world through a different paradigm. This paradigm includes a spiritual dimension, although not one attached to any given religion, and one that includes the subtle unseen forces. It does not always sit comfortably with conventional modern science. However, we can now see with hindsight that biodynamic principles and practice have produced startling insights into

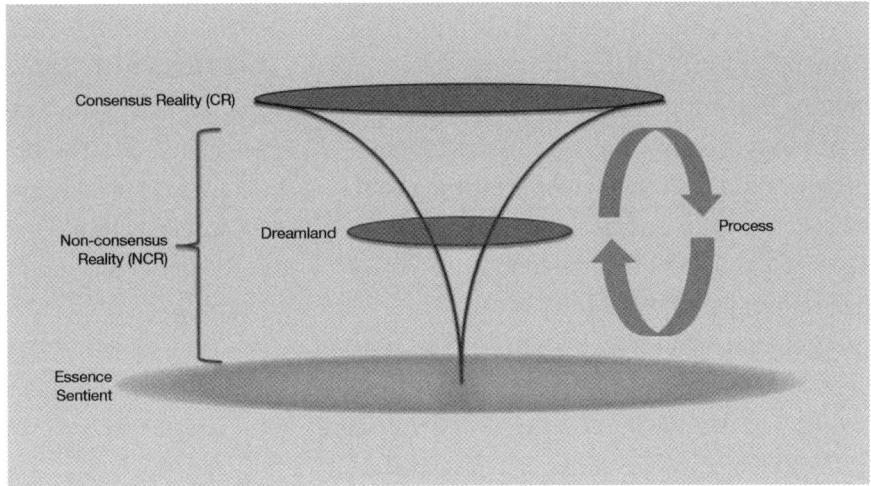

Figure 3.2: Mindell's model for different states of awareness.

how to produce food sustainably, that go way ahead of current scientific understanding. It has served as a catalyst for other regenerative forms of agriculture. Steiner used Goethe's qualitative research methods to develop a profound, holistic and practical grasp of what makes soil healthy and fertile.[10]

Arnold Mindell describes the world view of many indigenous peoples, and other spiritually inclined people, as a world view or reality that occupies two different states at the same time. He speaks of a 'consensus reality' world and the 'dreaming' world (Figure 3.2). It is psychologically healthy and indeed necessary to inhabit and move between these two states all the time. If we don't, we become mentally ill. We sleep, we dream, we process thoughts and events, we create, we do spiritual practice, we make and enjoy music and art and we love being in nature. All of these things require us to access the 'dreaming' or the non-consensus reality that we exist in.[11]

Darrell Addison Posey describes the indigenous view of agricultural practice as one in which farmers:

…generally view this knowledge as emanating from a spiritual base. All creation is sacred, and the sacred and secular are inseparable… in this sense, a dimension of traditional knowledge is not local knowledge, but knowledge of the universal as expressed in the local … thus knowledge of the environment depends not only on the relationship between humans and nature, but also between the visible and the invisible spirit world.[12]

Julia Wright, an associate professor at Coventry University, describes this as 'quantum-based agriculture', that is, as working with 'quantum and consciousness principles and practices', which involve the perception of invisible subtle energies. In biodynamic farming, this means working with the subtle rhythms of the earth, the moon and the stars and observing what effect they have on plant and animal growth. Such practices have arisen all around the globe in different agroecological practices and different indigenous world views, including in pre-industrial farming systems in Europe.[13]

Drawing on these perspectives, biodynamic farmers explicitly inhabit the consensus reality and the dreaming worlds, or visible and invisible worlds, at the same time, much as indigenous people might. Thus biodynamic farming could be considered an early form of European agroecology, except that the term 'agroecology' had not yet been coined in 1924.

Steiner's eight lectures gave rise to a set of principles and practices that the farmers could try out and develop into a farming system. Some of their elements were ancient and established; others were new. Some similarities can be seen with Vedic farming methods (see Chapter 4).

Steiner's philosophical approach to farming and food has inspired some of the most radical rethinks of modern sustainable food systems. Key to all this is the 'threefold path' that suggests that farms and farmers need to work on an environmental, a social and an economic level at the same time. From this has come a great deal of the creative thinking for modern sustainable farming systems.

Principles of biodynamic food production systems

The 'Demeter standards' list the principles that a biodynamic farm should follow. Steiner did not list a set of principles. They were developed subsequently by biodynamic practitioners and in particular Pfeiffer and the experimental circle at Dornach.

A biodynamic farm should:

1. become an organism with closed loop systems.
2. work with nature rather than against it.
3. create biological cycles and avoid pollution.
4. grow and produce high-quality food that is sold as locally as possible.
5. connect plants and animals and people to cosmic and planetary movements and work with them.
6. promote biodiversity from the soil up (10 per cent of each farm should be given over to biodiversity).
7. be socially and community connected.
8. be economically viable and provide right livelihoods.
9. allow animals on the farm to express their nature to ensure high welfare.
10. enhance the landscape.

Brief explanations of what these principles mean are listed below: [14]

1. **The farm as an organism**

   The key principle of a biodynamic farm is that it is perceived as an 'organism' in its own right, in which is developed a closed loop system. Steiner said, 'a farm comes closest to its own essence when it can be conceived of as a kind of independent individuality, a self contained entity'.[15]

   The concept of a farm as an organism implies that it will obtain most of its inputs and use up most of its waste products on the farm, selling the food that it produces to the community. 'Closed loop system' is an ecological term for a dynamic stable ecosystem, such as a woodland, where all the outputs are recycled as inputs. A circular food economy is one where all the waste streams from the food systems become the inputs back into the farms; the gluts and waste food are either fed to livestock or composted.

   How, then, does a farm sell excess food without bringing in extra nutrients from outside the system? The answer is that it is sunlight that provides energy to the system via photosynthesis, and the careful management of animal manure and legume crops that give fertility to the system. Through 'closing the farm gate' to outside inputs the farm develops in its food a 'terroir', or particular taste, dependent upon the soil, the climate and the production method. The produce becomes 'of the place', of the soil and the climate – truly local food. Its taste has a unique character. We are accustomed to this concept in wine and cheese, but not fruit, vegetables, grain or eggs.

   It is difficult to create a fully closed loop farming system. Steiner was aware of this.

   He suggested that farmers move towards this aim, even if they might not achieve it completely.

2. **Work with nature rather than against it**

   If a problem arises on a farm, the farmer will take time to observe the problem, and view it as an imbalance in the system that needs rectifying. A problem will be solved through working with rather than against nature. This can be illustrated with the approach to pests and diseases on a biodynamic farm. Rather than tackling a pest or disease problem once it has occurred, the biodynamic farmer aims to increase the health of the system so that the pest or disease does not arise in the first place. This is accomplished mainly through good soil management, correct growing conditions, choosing resistant crop varieties, choosing resilient animal breeds, and the right stocking level for animals. A biodynamic farmer will create habitat for predatory insects that will eat crop pests naturally. This is an instance of 'functional biodiversity'.

   The biodynamic movement has led the way in producing and supporting seed houses.

   These seed companies produce open-pollinated seeds that are themselves grown in biodynamic systems. These varieties flourish on biodynamic farms; they have not been bred to need lots of nitrate fertiliser, and tend to be genetically resistant to diseases. They are the polar opposite of GMOs and gene-edited seeds that are patented and sold for profit. In the UK the Seed Co-operative is a registered biodynamic seed house, not for profit, selling commercial quantities of outbred vegetable varieties.[16] In Germany, Bingenheimer collates seeds from 88 producers and sells seeds to 45,000 different customers.[17]

3. **Create biological cycles and avoid pollution**
Biodynamic farms have pioneered and developed composting systems to return to the soil the waste products from crops and animals. The compost combines with the soil particles to create humus, which stabilises the soil structure and forms soil aggregates. It increases the soil's organic matter, water-holding capacity, fertility, air content and cation exchange capacity (CEC), the latter reflecting the soil's ability to hold on to nutrients. The increase in the soil's microbial content leads to increases in the earthworm population, plant growth and crops' mineral content. The nutrient-dense food arguably tastes better and is more nutritious.

The increased microbial content will also prevent the leaching of nitrates from the soil, thereby preventing the eutrophication of local waterways. The higher level of organic matter in the soil also increases carbon sequestration. It has been found that biodynamic soils sequester 25 per cent more carbon than organic soils.[18]

Biodynamic farming communities pioneered the concept and practice of the reed bed and pond sewage systems. These treat raw sewage from rural communities using biological systems, discharging clean water back into the environment. In the UK a company called Ebb and Flow pioneered these biological sewage systems; they are beautiful, full of wildlife and very effective at cleaning black and grey water.[19]

4. **Grow and produce high-quality food that is sold as locally as possible**
Biodynamic farms are often mixed farms, producing a range of products that are sold directly to the consumer via the CSA (community-supported agriculture) model, farm shops or farmers' markets. The Demeter symbol has become synonymous with high-quality food, health and taste. Biodynamic wines in particular are highly prized.[20] It was from the community-connected farm principle that Trauger Groh developed the concept of CSA in the 1980s at his Wilton Farm in New Hampshire. His concept was that the community of customers or consumers needed to commit a regular amount of money every month to share the risk and surplus of farming, in order to ease the financial stress upon the farmer. The 'community' in effect buy from the farmer the entire crop or expected yield of meat or milk before it is produced; in return they receive a weekly share of the yield. The food is often sold as raw as possible.

Many biodynamic farms sell raw milk directly to their customers. From this model came the concept of the weekly box of vegetables, meat or milk which is now widespread in Europe and the US. Box schemes and CSA are now widely used across the organic and biodynamic movement.[21] (See Chapter 11 for a fuller description of alternative trading systems.)

5. **The farm is cosmically connected**

To connect a farm and food to the cosmos means that the work on the farm is done to the rhythm of the moon, planets and constellations. The section on 'Practices of biodynamic food production systems' explains how this is done. The food is believed to become more 'vital' or alive and full of forces from the moon, planets and constellations if it is grown to their rhythms. Working to the biodynamic calendar acts as a form of research and nature connection for the people on the farm as they become more aware of how these subtle forces affect plant growth and animal behaviour.[22]

6. **Foster biodiversity from the soil up**

Each biodynamic farm is required to commit 10 per cent of the farm to supporting biodiversity. Two biodynamic farmers, Marjorie Spock and Mary Richards, are unsung heroines of the modern environmental movement. They had studied biodynamic farming and Steiner's anthroposophy in the 1920s in Switzerland. They subsequently set up a small biodynamic farm on Long Island. In the 1950s the US federal government sprayed 3 million hectares of forest and heavily populated land with DDT, an organochlorine pesticide mixed with kerosene, in an attempt to halt the spread of the gypsy moth. This was devastating to the forests of upstate New York. Spock and Richards took out an injunction against the federal government in 1957 to try to stop the spraying. When the spraying went ahead anyway, they sued for damages. Because Richards suffered from 'multiple chemical sensitivity' they grew and cultivated their own food using the biodynamic methods. Pfeiffer was a witness in the trials and wrote a long research paper on DDT and the damage it does. They lost the battle in the federal court with the US government. The documentation and research showed the dramatic effect of the DDT on the bird population on their farm. All of this information was given to Rachel Carson and formed the basis for *Silent Spring*, published in 1962 and still in print today. This book is credited with starting the modern environmental movement as we know it today. Needless to say, the spraying of DDT did not solve the gypsy moth problem.[23]

7. **The farm is socially connected**

Since less than one per cent of the population of the global north work in agriculture, most industrial farms are bereft of people and are disconnected from the community. Biodynamic farms strive to be socially connected by including people in their activities. This can be through supplying their customers with seasonal food and inviting them to help with some of the large jobs on the farm such as the potato harvest. It can take the form of therapeutic work. The care farm movement sprang out of the biodynamic movement and the Camphill movement.

8. **Farms should be economically viable and provide right livelihoods**

Farming is capital rich and needs a lot of investment to make it work. Farmland is so expensive in many parts of the global north that it is almost impossible to start farming without either inheriting a farm or having large amounts of money. Tenancies are available, but the biodynamic farming system takes three to seven years to become fully established, which makes short-term tenancies problematic. The Biodynamic Land Trust was set up in 2011 to address this issue. It pioneered the concept of a 'community-owned and -connected farm'. The trust buys a farm and holds it in perpetuity for sustainable farming. Through buying shares in the Biodynamic Land Trust, people in the community can obtain a real stake in a farm and its development. Shareholders receive no financial return on their investment, but may benefit from living next door, having access to the food from the farm, or gaining a means to support carbon sequestration and biodiversity. Entrant farmers on the trust's farms are given long-term tenancies and know that their work developing the farm will be continued by the next generation after them.[26] A number of farmland trusts have now been set up for this purpose across the world – for example, Yggdrasil in Burlington, Wisconsin.[27] There are extensive German biodynamic farmland trusts.

9. **Allow animals on the farm to express their nature to ensure high welfare standards**

Biodynamic farms are by their nature mixed; animals and animal manure are an important part of the fertility system. The presence of cows on the farm is key. The welfare of the animals on the farm is of the highest standard, with low stocking rates, suitable breeds for the climate and soil, and the aim of producing the animals' food on the farm so that they have a healthy diet. The cows are left with their horns and are pasture fed. Cows in most farming systems are 'debudded' or have their horns burned off when young. The chickens are left with their beaks intact and allowed to feed outside all year round. Chickens are regularly debeaked in industrial systems because they peck each other in stressful conditions. Local abattoirs are used for slaughtering animals from biodynamic farms, so they do not become stressed or unwell through travelling long distances. The animals do not require the high levels of antibiotics that animals in the industrial systems need. Steiner predicted in his lectures that feeding cows with cattle feed made from dead cattle would create health problems both for the cattle and for people, as we found out with the CJD and BSE outbreak in the 1990s.[28]

10. **Enhance the landscape**

A biodynamic farm should be beautiful. Andy Warhol said, 'I think having land and not ruining it is the most beautiful art that anyone could ever want to own.' Biodynamic farms enliven and enrich the landscape. Running them is a creative process.'[29]

Care farming

Camphill Village Trust

The Camphill organisation was founded in 1938 by Dr Karl Koenig. He was a political refugee from Austria who arrived in the UK just before the outbreak of World War II with a group of young helpers and a plan to set up residential care and a school for children with learning disabilities. He was given land near Aberdeen on the Camphill Estate. At this time, children born with learning disabilities were regarded as ill. Koenig's approach was one of 'curative education': giving these young people an education and a meaningful life. Camphill communities have since spread across the world and today take only adults with learning disabilities. In the UK, there are nine of them, offering a home to a total of 400 adults.

Residents are usually based on a farm, with small housing units where life is lived as a family, and they have the opportunity to work on the farm or in the craft studios in the Camphill community.[24]

Ruskin Mill

Ruskin Mill Trust is an independent specialist training provider for young adults. It offers residential and day care and education in over 13 centres across the UK. Aonghus Gordon is the CEO and has pioneered these exemplary centres for young people to engage with the world. His vision is that all young adults need to be engaged with 'heart, head and hands'.

They work on farms and in craft workshops and 'are given the tools to transform material' and so transform themselves. Ruskin Mill's work with young people is inspired by anthroposophy, John Ruskin and William Morris. The farms are biodynamic and the buildings on their sites are beautiful and offer an extraordinary creative environment for these young adults. The trust offers training to staff and interns in biodynamic farming and gardening and in therapeutic work.[25]

Practices of biodynamic systems

What in practice do biodynamic farms do to implement these principles that make them distinct from other farming systems? First of all, biodynamic farmers think differently. They consciously think about how to adhere to and implement the principles outlined above. They consider how to produce economic, environmental and social yields, how to achieve multiple yields, rather than just the one, economic yield that is characteristic of industrial farming.

Biodynamic farming was the first ecological form of farming to arise in response to the introduction of industrial farming systems. It was the precursor of organic and agroecological farming systems and, as such, shares many similar practices with them.

However, it differs from organic and agroecological systems in two particular practices: the use of biodynamic preparations and the use of the biodynamic calendar.

The biodynamic preparations

Steiner recommended the use of seven preparations, as came to him in his spiritual contemplation. To the industrially trained farmer, to the scientifically trained mind, they are the most intellectually challenging of all the practices of biodynamic farming, but it is interesting to note that biodynamic farming has a high uptake in countries like India where spiritual practice and farming practice are regularly entwined. This section will explain what the preparations are and then how modern scientific understanding is catching up with this practice. The section 'Do biodynamic food production systems work?' will also examine the scientific evidence for the effects these practices have on the soil and the food that is produced using them.

I sometimes describe biodynamic farming as 95 per cent science and five per cent magic. This helps many people overcome the hesitation they might have in engaging with these practices and using the preparations. Most of us like to live in the consensus reality world but know that we need a little bit of magic to enhance our lives. The preparations and the calendar constitute a way in which to connect to the farm and nature on a deeper, more spiritual level; to use Mindell's words we can describe this as the 'dreaming level' of the farm and the land. Having taught these methods to hundreds of people over the years, I have found that when people actually take part in the stirring, in making the compost and in applying the preparations to the soil, the crops and the compost heaps, they find that these procedures all feel very natural. They bring a sociable and/or meditative aspect to farm work, which can be relentless and solitary for much

of time. I like to invite others to join me for the stirring. With a cup of tea and some food, it gives a pause to the farming activity at the pivotal times of the year, namely, the equinoxes.

The seven biodynamic preparations are added to the soil, plants and compost heaps on the farm. The formal names for the preparations are listed below, but the preparations are also often referred to by a number, given in brackets. This is simply the number they were allocated in a catalogue many years ago. The preparations can be bought ready-made from the regional Biodynamic Association or they can be made by the farmer and farm community.

- **Cow horn manure (500)**
  Fresh cow manure is collected in the autumn, placed in a cow's horn, buried at the autumn equinox for six months and dug up at the spring equinox (Figure 3.3). The preparation is used in small quantities. Although the amount is small, it is considerably more than a homeopathic dose. The cow horn manure is added to water, stirred for an hour in alternating directions to produce an 'alternating vortex' and then sprinkled over the soil in the autumn and winter months, in the afternoon or evening (Figure 3.4). One unit is 20ml, placed in 10–15 litres of water, to be spread over 0.4 hectares. The use of horn manure over a period of time increases soil biodiversity, increases the amount of humus in the soil, revitalises the soil, encourages root growth and makes the soil responsive to the cosmic forces. It has to be done at least once per year, preferably twice, in the autumn and then again in the spring.[30]

- **Horn silica preparation (501)**

  Ground-up quartz or silica is mixed with water to make a paste and packed into a cow's horn and buried at the spring equinox over the summer period. It is dug up at the autumn equinox. It is stored in a glass jar in the sunlight. This preparation is used in small quantities. One unit is one ml placed in 10–15 litres of water sprayed over 0.4 hectares. The horn silica is added to water and stirred for an hour with an alternating vortex in either a barrel or a bucket, depending on what quantity is required. It is sprayed as a fine mist on crop plants as they start to reach a stage of maturity, such as when lettuce hearts up, or apples begin to form in the summer months. It is applied in the morning. Ideally, it would be applied on the appropriate calendar day as described below.

  This preparation adds microbial activity to the leaves of the plant, regulates the metabolism of the plant, aids ripening and reduces fungal diseases, increasing the keeping quality, flavour and nutritional density of the crop. Animals eating fodder that has been sprayed with horn silica show better health as well. The horn silica is about light and ripening the crop; it helps crops and animals to be resistant to fungal diseases and pest attacks. The horn manure on the other hand is about fueling the growth

Figures 3.3 and 3.4: Horn manure. Above: Fresh cow manure is buried for six months. Below: A solution mixed in water is sprinkled over the soil in autumn and winter months.

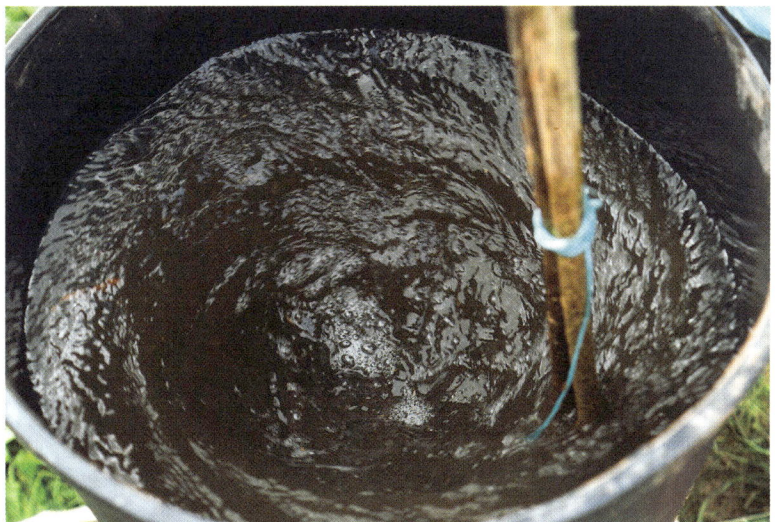

of the plant and bringing moisture to the soil. One is applied in the morning, the other in the afternoon; one in the winter and the other in the summer. They are opposite, yet both necessary to balance out plant growth and animal and human health. A little like the concept of yin and yang, they need to be in balance.[31]

The compost preparations

These flowers, leaves and bark are collected, wrapped in animal 'sheaths' and buried at different times of the year. About a dessertspoonful is added to a compost heap of 1m³ in size. Their function is to balance the nutrients, aid the rotting process and add vitality and microbial activity to the heap.

The compost-making process is similar to that found in most farming or gardening books, but the important thing to remember is that the biodynamic form of composting can be thought of more as a fermenting process. It is important that the biodynamic compost heap is not overheated, destroying all of the microbial activity, but is kept at a temperature that allows the microbes to flourish. The temperature is specifically 55°C. For more detailed descriptions of compost-making please refer to *A Biodynamic Manual* by Pierre Masson.

Each plant that is used as a compost preparation is also a traditional medicinal herbal remedy. Each one is linked, in old medicinal almanacs such as *Culpeper's Complete Herbal*, with a different planet, as are the animal sheaths. Each animal sheath is linked to a different metabolic process in the animal. In a way they are healing medicinal cures for the land and are

healing to the metabolic processes occurring on a farm. Steiner also suggests that they sensitise the soil and plants to planetary forces. The plants and sheaths that have been chosen are linked to the array of planets in the solar system, so bringing in the full array of planetary forces to the compost heap, the soil and the farm. Compost-making is crucial to the biodynamic farm; large mixed compost heaps built with the biodynamic preparations help the heap to break down quickly and efficiently. This compost is then spread over the farm and contributes to building the soil ecosystem. If it is a large farm and there is not enough compost to spread all over the area, the compost preparations can be incorporated into the horn manure preparation in the form of composite mixtures.

The compost preparations plants and sheaths are as follows:

- **Yarrow flowers (502)** – packed into a stag's bladder and hung in the light and then buried in the soil for six months. In the heap this regulates the balance between nitrogen and potassium processes. The yarrow plant is used for staunching wounds and stopping bleeding. It is also used as a divination plant in China for consulting the *I Ching*. The yarrow is placed in a bladder, which is part of the urinary system that cleans toxins out of the blood. Both yarrow and the bladder are related to the planet Venus.

- **Camomile flowers (503)** – packed in a cow's large intestine buried in the soil for six months. This regulates and stabilises the calcium and nitrate processes and helps with the 'digestion' process in the heap. Camomile is a herb used to aid digestion, and it is placed in the organ that does the digesting, the colon. Camomile is related to the sun. (Strictly speaking, the sun is not a planet but in this context it is treated as one.)

- **Stinging nettle (504)** – buried in the soil for 12 months. This regulates the iron and the nitrogen processes in the heap and the formation of humus. It organises the compost heap. The nettle is associated with blood and is used as a herbal tonic to improve iron and blood circulation. Both nettle and arteries are related to the planet Mars.

- **Oak bark (505)** – packed into a sheep's or cow's skull and buried for six months. This has a relationship to the calcium in the soil and helps prevent plant fungal diseases. The oak bark is high in tannins. The skull is home to the brain and the nervous system in animals. Both are related to the moon (although the oak tree itself is related to Mars).

- **Dandelion flowers (506)** – wrapped into the mesentery of a cow buried for six months. This regulates the silica and potassium in the heap. The dandelion is used as a diuretic. The mesentery holds the main organs in place: it envelops the peritoneum and the pancreas, which organises the metabolic systems. Both the dandelion and the liver are related to Jupiter.

- **Valerian flowers (507)** are soaked in water, filtered and put in a glass jar in the sunshine.

The valerian acts as a warm blanket to the heap, helping it to warm up, and it regulates phosphorus. Valerian is used as a sedative and a sleeping herb and is related to the planet Mercury.

This information is compiled from lots of sources. It represents a deep dive into ancient paradigms of medicine and European culture. Understanding these plants and their uses both medicinally and culturally is a journey; this field of study is called 'ethnobotany'. To explore the historical and current use of the plants in these preparations, I would recommend reading either Sattler and Wistinghausen or Masson.[32] Bockemuhl and Jarvinen explore the plants and sheaths from a Goethean angle.[33] *Culpeper's Complete Herbal* and Richard Mabey's *Flora Britannica* offer some further context for the ethnobotany of these plants.[34] Jonathan Code's *Muck and Mind* provides a fascinating insight into these preparations from a Goethean and alchemical perspective.[35]

Once these preparations are dug up they are stored in terracotta pots plunged into damp peat in a cool dark area. They are then used in small quantities in the compost heaps as the heaps are made.

Maye Bruce used the same plants mixed with honey as the basis for her Quick Return compost starter. The Maye Bruce or 'QR' preparations are made without the animal sheaths and are readily available today as compost activators bought as a powder in a packet and labeled as QR compost activators. Bruce was also one of the founders of the organic movement in the UK in the 1920s.[36]

The biodynamic calendar

The second unique practice that biodynamic farmers use is working around the rhythms of the moon, the constellations, and the planets. The moon has four rhythms in its orbit around the earth. Two of these are described here.

The first rhythm of the moon is the synodic cycle. This is the one that most people are familiar with as the moon waxes and wanes. It is well known that sowing seeds a few days before a full moon is known to improve germination.

Figure 3.5: Cover and sample page from The Maria Thun Biodynamic Calendar.

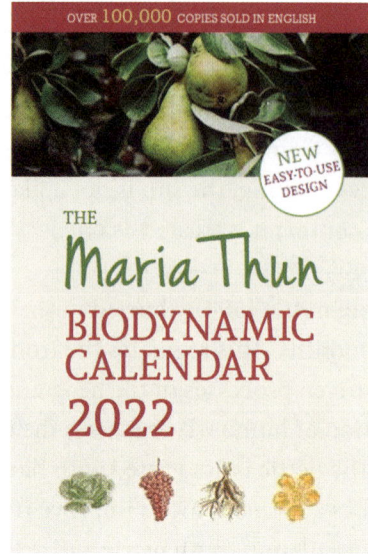

The second one is the sidereal or 'star' calendar. It's an ancient way of working with the land and crops and is still in use in many parts of the world today. The information is found in almanacs still widely available across Europe and North America. In England in 1879 Richard Jeffries wrote, 'the cottage astrologer has a whole table of quarters, aspects and so on, and lays much stress upon the day and the hour of the change; indeed, it is a very complicated business to understand the moon'.[37]

The practice of using the sidereal calendar observes when the moon is 'in front' of the constellations that make up the signs of the zodiac. We commonly read our 'stars' and we know that the sun takes a year to pass in front of all 12 constellations or signs of the zodiac. The sign of the zodiac in which the sun stands is thought to influence our lives and moods.

The moon takes one month to rotate through the signs of the zodiac and these are divided into four groups: water, air, fire and earth. In the sidereal system of growing, when the moon is in a water sign it influences leaf crops, when in an earth sign it influences root crops, when in an air sign it influences flower crops and when in a fire sign it influences fruit crops. This system has been in use for 2,000 years, maybe more, and in my experience farmers will only continue a practice if it has some value. Steiner would have grown up in his native Croatia with the local farms regularly using these almanacs.

It is the sidereal rhythm that the biodynamic farmer uses to plan when to apply the preparations. The horn silica preparation is thought to be most effective on the developing plant if it is applied on the right calendar day. The horn silica should be sprayed on a young apple crop on a fruit day. To hearten up lettuce it should be applied on a leaf day. The horn manure is applied on an earth day. Some farmers and growers also organise their cropping and seed sowing so that all crops are cultivated on the relevant day. This is very complex to achieve in practice, however, when the vagaries of the weather are factored in.

The biodynamic calendars (Figure 3.5) are produced every year by Floris Books and are available to buy from the Biodynamic Agriculture Association in the UK and many other places.[38]

Do biodynamic food production systems work?

Sceptical readers may well have read the description of biodynamic farming and felt challenged or dismissed it out of hand. Biodynamic farming works at many levels. Primarily, it is the quality of the food that stands out. Secondly, scientific research proves the changes in soil health that arise from long-term biodynamic practices.

Regarding the quality of the food, 'the proof is in the pudding', or in the following case study the wine. Biodynamic wines are among the best wines in the world and demand a high price. Many vineyards convert to the biodynamic system purely because of the quality of wine that is produced. Professional wine tasters in large supermarket chains use Maria Thun's calendar to taste the wine on fruit days. This gives them a clearer sense of the full array of flavours in the wine.

Biodynamic wine

*Monty Waldin made biodynamic wine famous in 2008 with his BBC documentary **Chateau Monty** on starting a biodynamic vineyard in France. He has written many books on the subject, and his website lists the world's biodynamic wines and wineries. He estimates that 4.5 per cent of all the vineyards in the world are either organic or biodynamic.[39] He says that 'if the best wines from Burgundy start to taste like the most banal wines from Australia because our soils have been made to become so similar, so denatured, from a saturation of the same chemical fertiliser and weedkillers, we will have lost not only our vinous heritage, but our raison d'être as winegrowers too'.*

The term 'terroir' is used in winemaking to explain the specific taste that wines take from the soil, the climate and the choice of grape variety. Biodynamic wine is stocked by most of the higher-end supermarkets in Europe; it can also be bought directly and shipped by producers straight from the farm.

An article in the Daily Telegraph stated, 'Earlier this month, Marks & Spencer held a 'fruit vs root' wine tasting – the first retailer to do so – to show the impact the moon can have on taste. At the end of the event, all but one of the critics correctly guessed which day was which.[40]

Biodynamic food is also regarded as of the highest quality and demands a high price. The entire produce of Fern Verrow Farm in Shropshire in the UK is bought up every year by Spring, a Michelin-starred restaurant in London.[41] Skye Gyngell says that the produce is 'fresh and vibrant'. Wendy Cook describes how biodynamic food has high flavour and good texture and will store for longer. She also explains how eating biodynamic food from a local farm means that you are more likely to eat seasonally, celebrate the cultural seasons of the food and eat with friends and family, all of which is very healthy.[42]

Dr Jens-Otto Andersen, a founding member of the International Association of Food Quality and Health and a biodynamic farmer, has taken this one step further. In his research in Denmark he has found that organic and biodynamic food contains higher levels of genetic repair enzymes. He has conducted experiments in which he cuts up organic, biodynamic and supermarket cucumbers (the latter grown in a hydroponic system) into thick slices, then wraps them in cling film and puts them in an incubator at 24°C for 14 days (Figure 3.6). He has found that, unsurprisingly, the supermarket cucumbers dissolve into mush, but many of the organic and biodynamic cucumbers grew back together again. To test the strength of these repairs he hung weights on the cucumbers. He found that the biodynamic cucumbers could support five kg before snapping and the organic ones two kg. His hypothesis to explain this is that the organic and biodynamic cucumbers have higher levels of genetic or DNA repair enzymes in them. This he conceptualises is a measure of the health and

With kind permission of Jens-Otto Andersen

Figure 3.6: The cucumber trials.

vitality in a plant, animal or person.

In the same way we need to re-define our concept of health, from absence of single diseases towards the ability of the organism to maintain a complex balance in the organs and the life processes during the major changes taking place recurrently during our life cycle.[43]

His suggestion is that eating organic and biodynamic food increases our ability to maintain our health as a dynamic process, one of constantly repairing and balancing out the biological processing happening in our bodies all the time.[44]

Secondly, the scientific evidence demonstrates how good biodynamic systems are at regenerating the soil

biodiversity. The 'DOK' trials were started in 1978 in Switzerland by the Swiss National Institute of Biological Agriculture (FiBL). Randomised plots were used to compare biodynamic (D for Demeter), organic (O) and conventional (K) (industrial) systems of agriculture over a long period of time. The researchers wanted to compare the outcomes of these different systems of farming in all of their complexity (Figure 3.7).

The use of randomised plot trials evens out the background variables in the complex systems – the variables in soils, sunshine hours and wind speeds – to facilitate the comparison of different farming practices. For instance, if the conventional plots had been allocated the best soil on the sunniest spot then you would expect those crops to have a higher yield. By having lots of small plots mixed up, the variables become distributed across all of the systems, so that meaningful measurements can be taken.

The scientists decided what to measure. What is key to this bit of research is that they had the foresight not just to measure the yield, as is so

Figure 3.7. Randomised plot trials.

With kind permission of FiBL.

common in modern agricultural trials; they also monitored the soil processes and the effect the systems had on the environment around them, or the 'multiple yields' of the systems.

They started farming on these plots using the standard methods that would be used in each farming system: the rotations, and the application of fertilisers, manure and preparations, according to the particular farming system.

The differences between the organic and biodynamic systems and the industrial system were evident after three years. But after 21 years the biodynamic soils had become significantly different from the organic soils. They became significantly more biodiverse, with more earthworms and beetles, more microbial activity, more microbial mass, more mycorrhizae, more weeds and better soil structure. The word 'significantly' is important here, since the data from these plots are analysed statistically, and this means that on all the plots for most of the time there is an increase in the diversity as well as the overall amount of soil microorganisms.

This work was written up and published in *Science* in 2002.[45] It is a very famous bit of research. What it proves is quite simply that biodynamic systems work on a physical level to create biodiverse and resilient soils. These trials showed that the industrial farming system had the highest yields, the organic and biodynamic systems having on average 20 per cent lower yields for some, but not all, crops.

In subsequent years these trials have gone on to show that biodynamic soils sequester 25 per cent more carbon than do organic soils, and use 30–50 per cent less energy than industrial farms per yield of crop. This means that one tonne of biodynamic wheat takes 30–50 per cent less energy to produce than the equivalent tonne of industrially grown wheat.[46]

I had the privilege at the time this research was published to be introduced to Urs Niggli, the Director of FiBL, by Martin Wolfe over dinner at the Soil Association Conference. As these trials were done by scientists and by farm workers who were not biodynamic farmers, I asked him what they had made of the results. How did they explain them? He told me that they didn't know. The science had discovered that this method worked but not how it had worked.

Dr Elaine Ingham is a soil microbiologist from the US who worked at the Rodale Institute in the US for many years, and for the US Department of Agriculture. She published her work on soil ecology, which she called the 'soil food web'. She found methods to isolate and to count the numbers of species of bacteria, fungi, nematodes, insects and worms in the soil. This is a huge task; it is estimated that there are still vast numbers of undiscovered species of bacteria and fungi in the soil. Ingham suggested that there are more living microbes in a teaspoon of living soil than there are people on the planet. It seems remarkable that no one had done this work earlier.[47] Ingham explained scientifically why the presence of soil biodiversity is key to a sustainable soil. The soil microflora, that is, the bacteria and fungi found in the soil, carry out many tasks. They extract nitrogen from the air and make it into nitrates available to plants to use. They mine phosphate and potassium from the clay particles and make them available to the plants for food. Some eat plant pathogens; they digest organic matter and more. They also die, and when they die they themselves become food for other bacteria and fungi. They build up the humus from themselves and from decaying organic matter. They are at the bottom of the food chain that supports the larger species: the nematodes, the worms, the beetles and the springtails. The nematodes, worms and beetles are in turn eaten by other beetles, larger worms, and birds and moles. All of these in turn die and are digested by yet other bacteria and fungi to become the nutrient soup that feeds the plants. What bacteria and fungi need to thrive is air (since they respire) and some food and some moisture. In many cases the food is the sugar that exudes from plant roots and the dead and decaying matter in the soil. All of these factors are found in the top 30 cm of well-structured, organically rich soils. We know from ecology that the more diverse an ecosystem is, the more resilient it is.

Ingham also discovered that any chemical-based fertiliser or pesticide applied to the soil will kill large amounts of bacteria and fungi, thereby impacting upon the 'web' and the resilience of the soil, in effect killing it.

From these two strands of research we can understand that biodynamic soils are the most biodiverse. This biodiversity is good for soil fertility, soil carbon sequestration and the building of soil organic content and water-holding capacity.

The horn manure may be part of the mechanism that enables the soil to become more biodiverse. The horn manure preparation is made from cow manure, and it is important that the cow is from a biodynamic or organic system, that is, has not had any antibiotics, and is ideally from a local farm. This manure is buried in a horn in the ground in the autumn, a cool time of the year. The manure undergoes a slow fermentation, breeding vast numbers of bacteria that originated from the cow's gut. When the cows horns are lifted, fungal mycelium can be seen growing through it. Small amounts of the horn manure are added to warm water and stirred for an hour, using the alternating vortex, this adds air to the mix, enabling the microbes to multiply.

Research on the horn manure has found that it is similar to other humic substances in its chemical makeup. Not only is it full of bacteria and fungi, but it is also made up of biolabile components that behave in a similar way to plant hormones. When applied to cress seedlings, the horn manure was found to promote growth in a similar way to the plant hormone auxin. The shape of the horn itself creates the conditions for fermentation rather than decomposition, so increasing the microbial activity. A specific group of fungi found in the cow manure reacts with the keratin of the horn to provide good conditions for the proliferation of these fungi. These then go on to seed the soil as a form of biocatalyst that helps all of the soil microorganisms to proliferate. These in turn foster the creation of soil aggregates in the soil.[48]

On a purely physical level, it may well be that the horn manure preparation behaves like an 'inoculation' of biodiversity across the farm, which when repeated over many years translates into a soil that is teeming with life. It's similar to repairing a gut biome through drinking kefir.

This could also explain how one of the farmers who asked Steiner to give them the lectures in 1924 could grow a monocrop successfully over a 10-year period without any build-up of pests or diseases. Ingham suggests that within a healthy soil antagonistic bacteria and fungi will outcompete the plant pathogens.[49]

Once a soil is teeming with life, it becomes easier to work and easier to grow food in.

The food will taste better because it will have a greater complement of minerals and secondary metabolites. This nutrient density imparts a 'terroir' or a taste that is specific to the soil and place. It will pass on health benefits to the livestock or people that eat it. The pest and disease incidence in crops and livestock will decrease because of the antagonists found in the soil. A soil that is full of biodiversity will behave like a farm immune system.

What does a biodynamic food production system look like?

Take a walk around a biodynamic farm and you will experience a wonderful, living, exciting place, full of wildlife and vibrant-looking crops, livestock and people. Many biodynamic farms are community-based farms, which make their produce available to people on lower incomes through work shares or through donations to food banks. The food tastes divine.

Biodynamic farms are found all over the world and are always worth a visit. The Living Farms project is traveling around the world to film biodynamic farms. The films are available on the Goetheanum website.[50] A third case study of a biodynamic farm is found in Chapter 10's description of Huxhams Cross Farm.

Case study:
Winter Green Farm, Eugene, Oregon

In 2017, I was fortunate enough to come across the Winter Green Farm stall at the weekly Eugene Farmers' Market, in Oregon. This is the farm's story.

Halfway down Route 126 between Eugene and the Pacific coast, not far from the legendary Route 101, is a beautiful valley with a beautiful farm. On our journey to this farm we took a wrong turning and ended up in a logging village; what we saw there was a scene of devastation. The hills were being clear felled. The place was full of heavy machinery. There were deep ruts in the soil and lots of bare soil, a huge industrial sawmill and massive trucks going in and out of the village, loaded with tree trunks. A few miles on, around a few more corners, we found Winter Green Farm, nestled in a long valley, serene and beautiful, the tops of the hills still covered in woodlands, and the valley full of crops, cows and sheep.

Winter Green Farm has 70 hectares of land in total in a mild and damp climate and produces a mixture of vegetables, berries, herbal plants, beef cattle and sheep, with tracts of oak forest and riparian and wetland areas given over to biodiversity. It is run by a community of six farmers, who state that their aim is to create 'a productive farm in harmony with the earth, humanity and ourselves'.

The farm was started by Jack Gray and Mary Jo Wade in 1980 as a form of homesteading and was put into organic conversion. In 1986 it went into biodynamic conversion. When I visited in 2017 they had created a community of six intergenerational farmers on the site so that the

Figure 3.8. Winter Green Farm, Eugene, Oregon.

farm could continue to evolve and thrive into the future.

The farm grows ten hectares of herbal medicine plants, namely the burdock root, making it the largest burdock root grower in the northern US. They grow ten hectares of vegetables and have 36 hectares of pasture grazed by sheep and beef cattle. The rest is given over to wildlife and biodiversity. They have a flock of 50 sheep and 26 beef cattle, whose calves are kept for two years before slaughter. All the animals are slaughtered on the farm. They practise mob grazing with the herds. (See Chapter 8 for an explanation of mob grazing.)

The vegetables are sold to 450–500 CSA customers in Eugene and Portland, and Winter Green Farm also attends weekly farmers' markets in Eugene and Portland, as well as selling some produce wholesale to restaurants and shops in the region. The vegetables are grown on a six-year rotation, three years of crops and three years of deep-rooting herbal leys grazed with cattle and sheep.

They make all of their own biodynamic preparations and compost on the farm, both of which go back into the farm system. These alongside green manure and grazing of the cows and sheep on the land in rotation with the vegetable crops provide all the fertility needed. At Winter Green Farm the preparations are stirred using a flow form and a mechanical pump.

The preparations are then applied with a tractor-mounted sprayer.

The farmers emphasise the fact that they need to make the farm financially secure so that it can continue to thrive and develop, and so that they can pay their staff a living wage and so that the farmers themselves can thrive too and contribute to the community they are in. The farmers contribute to their extended farm community by mentoring staff, hosting conferences and farm walks, hosting visits and celebrations and just by having enough time to enjoy the farm they live on without being completely exhausted (see Figure 3.8).

They also donate food to food banks so that people on a lower income can access their beautiful food and they work hard to provide events on the farm so people can visit and enjoy and connect to their food. Their concept of 'food justice' says it all:

Growing great food and making it available to people, regardless of income, is one of our core values. We have developed a close link with our local food bank, and through the combined effort from our labour and donation from the community we have provided thousands of pounds of produce to those in need. We also seek to pay our workers a fair wage while creating a work atmosphere of mutual respect, learning and community at our farm.[51]

Case study: Sekem, Egypt

In the midst of sand and desert I see myself standing at a well drawing water. Carefully I plant trees, herbs and flowers and wet their roots with precious drops. The cool wet water attracts human beings to refresh and quicken themselves. Trees give shade, the land turns green, fragrant flowers bloom, insects, birds and butterflies show their devotion to god the creator, as if they were citing the first sura of the Qu'ran.

Dr Ibrahim Abouleish (1937–2017) wrote these words about the founding of Sekem in the desert 60 km north of Cairo. Abouleish was born in Cairo in 1937, son of a soap merchant. He moved to Austria to go to university and married and had children there. While studying for a doctorate in pharmaceuticals he came across the work of Steiner and decided to move back to Egypt to start a farm in 1977. He bought a 70-hectare plot of land on the edge of the Nile Delta but it was pure desert. In *Sekem* he describes how he and engineers dug wells, planted trees and built the first house on the farm – the round house.[52] From that beginning he brought the soil on the farm back to life by bringing in cattle and producing compost. They grow cotton and herbal plants for teas and medicines. They branched out to other farms and helped other farmers to convert to organic or biodynamic systems to grow vegetables for export to Europe and for home consumption. In his book Abouleish describes how the aerial spraying of the cotton fields in Egypt caused pesticides to drift on to his cotton. To counter that, he worked with academies and researchers to find biological and cultural controls for the pests on the cotton plants, which they accomplished after three years. These were then rolled out across Egypt to reduce the use of pesticide sprays.

The farm grew to include kindergarten, primary and secondary schools, technical training and finally a university. They have developed mutually beneficial trading between the farms in Egypt and markets in Europe which enables a fair price to be paid to the farmer.

The various companies include Isis, selling fresh vegetables, honey, spices, dates and oils; Atos, selling herbal medicines; Naturetex, selling cotton and fabrics. Plus more. Abouleish died in 2017 and the CEO is now his son Helmy. Helmy has taken the vision further with detailed development plans up until 2057, in which they aim to roll out sustainable development in Egypt with more biodynamic and organic food production in the desert. They have developed the business along the lines of the threefold intersections of ecology, culture and economics. They call this the 'economy of love':

Sustainable development towards a future where every human being can unfold his or her individual potential; where mankind is living together in social forms reflecting human dignity; and where all economic activity is conducted in accordance with ecological and ethical principles.[53]

The family of businesses that make up Sekem are all flourishing. Sekem and its various companies now occupy 2,100 hectares; they have planted 60,000 trees, employ 2,000 people and have sequestered 0.5 million tonnes of carbon dioxide. Sekem and Abouleish were awarded the Right Livelihood Award, otherwise known as the 'alternative Nobel Prize', in 2003. In 2015 the UN awarded them the Land for Life award for their work against desertification.

The contrast between the farm in 1977 and how it is now is astonishing (see Figure 3.9). They have an eco-village for visitors to visit, or stay in while they explore the farm, or you can watch the film on their website to visit from your armchair.[54]

Figure 3.9: Sekem, showing the original roundhouse (top) contrasted with how it is today (bottom).

Training in biodynamic food production

Biodynamic training is available around the world and listed on the Goetheanum Agricultural Section website and is currently being unified by the Agricultural Section at the Goetheanum.[55] Many different levels of training are available. There are many short introduction courses of one to four days. (We run one over four days at Huxhams Cross Farm.) These are normally aimed at interested gardeners.

In the UK there are one- to two-year internships at level three (further education) for young people training in professional farming and horticulture, with placements on a farm and block courses of theory in the winter months.

There are courses for existing farmers and horticulturalists wanting to convert their business to biodynamic methods. In Germany farmers are invited to attend a six-day introductory course in the biodynamic systems as they consider entering into the biodynamic conversion process. There is also CPD for farmers in biodynamic preparation-making, viticulture and beekeeping. In the Netherlands there is a 10-week course in biodynamic farm management, run over two years in Warmonderhof, which is state funded. For existing biodynamic farmers there are farmer-to-farmer groups in the form of experimental circles to advance their understanding.

An online learning course is available with the Biodynamic College in the UK, and webinars are available on the US Biodynamic association website.[56] All of this training is listed on the Goetheanum website.[57]

Where is the biodynamic food production movement going?

The biodynamic farming movement is the oldest of all the sustainable farming systems and will be 100 years old in 2024. So why does it have such a low uptake in comparison with organic farming, agroecology or regenerative agriculture? Biodynamic farming has been a precursor and catalyst in the development of organic farming, the environmental movement, CSA, community-owned farms, farmer-to-farmer study groups and the therapeutic use of farms. Many of these methods are adopted widely. Why not biodynamic farming itself?[58]

One of the main barriers to uptake is the use of the preparations, which many find intellectually challenging. The organic pioneer Lawrence Hills famously called biodynamics 'muck and magic'. However, although it may give some rationalist farmers a headache, biodynamic vineyards have clearly shown that the practice of putting on the preparations on a large scale is itself relatively straightforward and brings great rewards in terms of the quality of the product, soil fertility and the price premium that can be gained. Biodynamics has helped transform rundown monocultural, eroding vineyards into beautiful, verdant, biodiverse, multicropping farms.

Another barrier to biodynamic farming is that the literature is difficult to understand. The original lectures by Steiner are almost impossible to understand unless they are read in a study group led by an informed biodynamic farmer or grower. Discussions and descriptions of the practice of biodynamic farming can be muddled, causing confusion. It is interesting that in India where spirituality and everyday life are more integrated, and there is a deep reverence for cows, there is much higher uptake. Meeting with biodynamic farmers, visiting their farms, hearing them talk about their practice, experiencing it with head, heart and hands are among the best ways to experience these systems.

So where is the movement going next? The Biodynamic Federation–Demeter International has set out a vision statement that says, 'It shall also be emphasised, that agriculture is seen as an essential foundation for both personal and societal development and that it will gain in importance as it provides solutions for all burning issues of the present including the economic, cultural, social, and ecological ones.'[59] Specifically, the vision:

- encourages humankind to take over the responsibility for the holistic development of the earth (ecology);
- impels and enables people to unfold their individual potential and develop their full consciousness (human development);
- produces wholesome and healthy food and other agricultural products that are of high quality and nourish body, soul and spirit (economic value creation);
- encourages people to live and work together in dignity, mutual respect and tolerance (social relationship);
- embraces the material and spiritual world and empowers humankind to be conscious of and embed the cosmic and terrestrial forces and substances (cosmic and spiritual impact).

What does this look like in practice? In the UK and the US the Biodynamic Associations are actively engaging with the public and other sectors of sustainable farming. They are building bridges to communicate the principles, practices and ethos of the biodynamic movement. In particular, they are integrating with other forms of the regenerative agriculture movement.

In the US this work is underway with a website packed with webinars, farmer training programmes and research portals. Links are being built with indigenous Native American farmers and growers and socially just food systems. The US Biodynamic Association positions itself as a part of the regenerative agriculture partnership. A recent study in the US suggests that consumers of sustainable food are increasing their own understanding of soil health and how it relates to carbon sequestration, the taste of the food and their own health. They are seeking out biodynamic food because it performs well in these areas.[60]

It is evident that biodynamic farming systems are among the most established and sustainable of all farming systems. As the need for these systems becomes more urgent in the next decade the biodynamic movement will need to disseminate their knowledge.

Information, training and expertise will need to be readily available and transferable for those who wish to add it to their toolkit. Thus the training on offer around the world is being joined up and made more readily available from valid trainers and training organisations. ■

In a nutshell

- Biodynamic farming and gardening is almost 100 years old.

- It was the first ecological farming system created.

- It was a precursor to organic farming.

- It aims to work towards a closed loop system creating a farm organism.

- It uses livestock manure and green manure to form composts for crop and animal nutrition.

- It relies on improving health as a means to reduce pests and diseases in the plants and animals.

- Pests and diseases are treated using biological or mechanical controls.

- It has very high welfare standards for animals.

- The soils and plants and compost heaps are treated with preparations two to three times per year.

- These increase the microbial activity in the soil, making these soils some of those best able to sequester carbon and most resilient in the face of climate change.

- A planting calendar is used to give seasonal rhythms to the farm work.

- Biodynamic produce is regarded as having the best taste and keeping quality by many connoisseurs, especially the wine. The food is nutrient dense, with additional health-giving properties from secondary metabolites.

- The concept of CSA (including box schemes) arose out of biodynamic farming.

- It has been a catalyst of much of the organic and environmental movement and introduced farmer-to-farmer peer-led learning groups of a kind now used across the world in agroecology systems.

- It is now considered part of the regenerative agriculture partnership.

Further reading, links and films

T. Petherick, *Biodynamics in Practice: Life on a Community-Owned Farm*, Sophia Books, Forest Row, 2010.

E. Pfeiffer, *Biodynamic Farming and Gardening*, Mercury Press, New York, 1938.

W.D. Storl, *Culture and Horticulture,* Biodynamic Farming and Gardening Association, San Francisco, 1979.

M. Waldin, *Biodynamic Gardening*, Dorling Kindersley, London, 2015.

Biodynamic Association UK (BDA). Their website offers book lists, research articles, back copies of *Star and Furrow*, and links to films. www.biodynamic.org.uk

Biodynamic Association US is a good source of free webinars, scientific papers and talks on biodynamic farming. www.biodynamics. com

International Biodynamic Association is based in Switzerland. www.ibda.ch

The Goetheanum Agricultural Section has a wide range of information on biodynamic farming, including many publications and links to farms around the world. https:// www.goetheanum.org/

Star and Furrow magazine is published twice per year in the UK. You can join the BDA and attend the biennial conference. There are also regional groups across the country.

Biodynamic College, in the UK, has an apprenticeship scheme and distance learning package. http://www.bdacollege.org.uk

Chateau Monty is a BBC series about biodynamic wine-growing. Available on YouTube. Visit Tablehurst Farm in the UK: https://www.biodynamic.org.uk/farm/#becoming

Visit Sekem Farm in Egypt: www.sekem.com/en/media/videos

One Cow, One Man, One Planet is a film about Peter Procter's work on biodynamic farming in India.

Symphony of the Soil is a film about soil ecology and Elaine Ingham's work. Available to buy or rent at https://www.symphonyofthesoil.com/

1   R. Steiner, *Agriculture*, trans. C.E. Creeger and M. Gardner, Biodynamic Farming and Gardening Association, Kimberton, PA, 1993, p. 89.

2   Ibid.

3   https://www.demeter.net/statistics

4   Steiner, *Agriculture*, p. 9.

5   P. Conford, *Origins of the Organic Movement*, Floris Books, Edinburgh, 2001.

6   J. Paull, 'The Betteshanger Summer School: Missing Link between Biodynamic Agriculture and Organic Farming', *Journal of Organic Systems* 6(2), 2011, pp. 13–26.

7   https://www.pfeiffercentre.org

8   Steiner, *Agriculture*, p. 187.

9   A. Mindell, *The Shaman's Body*, HarperCollins, San Francisco, 1993.

10   M. Bruce, *Common Sense Compost Making*, Faber & Faber, London, 1946.

11   Mindell, *The Shaman's Body*.

12   J. Wright, 'Quantum-Based Agriculture: The Final Frontier', Coventry University, 2017.

13   D. Posey, *Cultural and Spiritual Values of Biodiversity*, UNEP, 1999.

14   https://www.bdcertification.org

15   Steiner, *Agriculture*, p. 27.

16   https://www.seedcooperative.org.uk

17   https://www.bingenheimersaatgut.de/

18   Soil Association, 'Soil Carbon and Organic Farming', 2009.

19   https://www.ebbandflowltd.co.uk

20   https://www.bdcertification.org

21   T. Groh and S. McFadden, *Farms of Tomorrow*, Biodynamic Farming and Gardening Association, Kimberton, PA, 1990.

22   Wright, 'Quantum-Based Agriculture'.

23   R. Carson, *Silent Spring*, Penguin Books, London, 1962; J. Paulli, 'The Rachel Carson Letters and the Making of *Silent Spring*', *SAGE Open* 3(3), 2013.

24   https://www.camphill.org.uk

25   https://www.rmt.org

26   https://www.biodynamiclandtrust.org.uk

27   https://www.yggdrasillandfoundation.org/

28   Steiner, *Agriculture*.

29   M. Zanardo et al., 'Metabarcoding Analysis of the Bacterial and Fungal Communities during the Maturation of the Preparation 500, Used in Biodynamic Agriculture, Suggests a Rational Link between Horn and Manure', *Star and Furrow* 135, 2021.

30   P. Masson, *A Biodynamic Manual*, Floris Books, Edinburgh, 2012

31   Ibid.

32   Ibid.; F. Sattler and E. Wistinghausen, *Biodynamic Farming Practice*, Cambridge University Press, Cambridge, 1992.

33   J. Bockemuhl and K. Jarvinen, *Extraordinary Plant Qualities for Biodynamics*, Floris Books, Edinburgh, 2006.

34   N. Culpeper, *Culpeper's Complete Herbal*, 1826; R. Mabey, *Flora Britannica*, Sinclair-Stevenson, London, 1996.

35   J. Code, *Muck and Mud*, Lindisfarne Books, 2014.

36   Bruce, *Common Sense Compost Making*.

37   R. Jeffries, *Wild Life in a Southern County*, Thomas Nelson, London, 1879.

38   M. Thun, *Biodynamic Calendar*, Floris Books, Edinburgh, 2020.

39   https://www.chateaumonty.com

40   The Telegraph, 'Biodynamic the New Organic', 28 September 2009.

41   J. Scotter and H. Astley, *Fern Verrow: A Year of Recipes from a Farm and its Kitchen*, Quadrille Press, London, 2015.

42   W. Cook, *Biodynamic Food and Cookbook*, Clairview Books, Forest Row, 2006.

43   R. Swann, 'Advances in Biodynamic Food Quality: The Cucumber Method', *Star and Furrow* 132, 2019, p. 7.

44   J.O. Andersen, *Vitality from Soil to Stomach*, Books on Demand, Norderstedt, Germany, 2019.

45   P. Mader, A. Fliessbach, D. Dubois, L. Gunst, P. Fried and U. Niggli, 'Soil Fertility and Biodiversity in Organic Farming', *Science* 296, 2002, pp. 1694–1697.

46   FiBL figures on energy used in organic farming: https://orgprints.org/id/eprint/37222

47   https://www.SoilFoodWeb.com

48   Zanardo et al., 'Metabarcoding Analysis of the Bacterial and Fungal Communities'.

49   R. Spaccini et al., 'Molecular Properties of a Fermented Manure Preparation Used as Field Spray in Biodynamic Agriculture', *Environmental Science and Pollution Research* 19, 2012, pp. 4214–4225; A. Morau and H.P. Piepho, 'Interactions between Abiotic Factors and the Bioactivity of Biodynamic Horn Manure on the Growth of Garden Cress (*Lepidium sativum* L.) in a Bioassay', *Chemical and Biological Technologies in Agriculture* 7(11), 2020.

50   https://www.livingfarms.net

51   https://www.wintergreenfarm.com

52   I. Abouleish, *Sekem*, Floris Books, Edinburgh, 2005.

53   https://www.sekem.com/wp-content/uploads/2017/05/SEKEM-Company-Profile-2017.pdf

54   Ibid.

55   https://www.sektion-landwirtschaft.org/en/thematic-areas/training

56   https://www.biodynamics.com/

57   https://www.sektion-landwirtschaft.org/arbeitsfelder/ausbildung

58   Wright, 'Quantum-Based Agriculture'.

59   https://www.ibda.ch

60   Hartman Group, 'Organic and Beyond', 2020.

Chapter 4

Organic food production

The health of the soil, plant, animal and man is one and indivisible.[1]

Eve Balfour

Introduction

The organic food production system was born soon after the introduction of industrial farming methods and arose out of detailed observations of ancient sustainable farming methods in China, India, Japan and Korea by European and American farmers and researchers. The organic movement arose subsequently out of collaborations and discussions in Europe and the US, starting as early as 1926.

This chapter will explain the principles and practices of organic farming, the long-term effect they have on the soil and the farming community, what kind of food they produce, and where you can find organic farms. We will explore how organic farming meets the four challenges of climate change mitigation and adaptation, biodiversity loss and producing enough food for everyone. Finally, there are case studies of successful organic farms, and signposts to where you can find out more about organic food production systems.

What are organic food production systems?

The first observations of farming systems, and the books about them, which evolved into the organic food production movement started as early as 1911. Since then organic food production has developed a clear methodology. The International Federation of Organic Movements (IFOAM), now known as Organics International, sets the standards and regulations for organic farming worldwide. These are administered by regional organic associations such as the Soil Association in the UK. Organic produce is sold under the trademark of the leaf symbol in Europe. The headquarters of IFOAM is in Germany.

Organic food is sold under a trademark all over the world and currently uses about 1.5 per cent of the world's farming land.[2]

Where did organic food production start?

Philip Conford, the historian of the organic movement, puts the start of organic farming in

1926 in the UK, although the Soil Association, the official body of the organic movement in the UK, was not formed until 1946.[3] As with biodynamic farming, this counterpart system of food production arose in response to the new chemical and industrial farming systems that were introduced from 1913 onwards.

A number of key figures brought the organic movement into being. To read about them all fully, I would refer you to Philip Conford's *The Origins of the Organic Movement*. But to simplify the story I have chosen nine of the founders and discussed the sources of their insights and ideas.

Organic farming principles and practices originated in China, Japan, Korea and India.

Key figures from the US and from the British Empire worked with and observed these principles and practices and then translated them for use in Europe, the US, Australia and Africa.

As early as 1911 Franklin King, an American soil scientist, published a book called *Farmers of Forty Centuries* based on his study tour of China, Japan and Korea.[4] King was the chief of the Soil Division of the United States State Department of Agriculture (USDA) at the time. He travelled to these countries in 1907 and studied their long-term sustainable methods of farming, built upon the concept of 'the rule of return'. Small-scale farmers in China composted their plant waste with animal and human waste, returning it to the soil as rich humus. They used green manures and legumes to build soil fertility and complex three-dimensional cropping systems to increase yields in small areas. King describes how rice paddies are also home to fish and ducks, and the edges are planted with trees such as plum or mulberry, with intensive vegetables grown nearby. He compared

the high yields of intensive food production in China with the situation in the US. He suggested that if this level of intensive but sustainable farming could be practised in parts of the US it would radically improve both soils and the amount of food the US could produce. This was 20 years before the Dust Bowl happened in the US, involving huge amounts of soil erosion and mass starvation and migration.

Sir Robert McCarrison was a British nutritionist, who chose to live in the Hunza Valley, in the foothills of the Himalayas, for seven years in the 1920s. Today the Hunza Valley is in the north of Pakistan; in the 1920s it was part of the British Empire and still part of India. He was intrigued by the Hunzas, a people who were fit and healthy. He noticed that the vitality they possessed was not just the absence of disease and included a robust strength and energy and a positive outlook on the world. He also noticed that the plants and the livestock too were 'vital' and rarely suffered from disease. After years of living in the Hunza Valley and then conducting experiments on rats, using the typical Western diet and the Hunza diet, he came to the conclusion that the vital health of the food and the Hunza peoples was due to their farming system. This farming system was based on humus and composting, as well as the type of food they ate: whole grains, lots of fresh vegetables and fruit, plenty of milk and butter and not much meat or alcohol. It was similar to the system that King observed in China.[5]

Sir Albert Howard was British, came from a farming background and started his career as an economic botanist in British India. There he met and married Gabrielle Matthaei, a gifted plant scientist, and they worked closely together. After Gabrielle's death he married her sister, and she too contributed to this work. They worked with Indian farmers to grow and breed plants that we can assume were for the benefit of the British Empire. Through this process he observed the traditional Vedic Indian farming system and realised that it produced plants that had no diseases and were very healthy. The Vedic methods of farming derived from ancient Hindu texts on how to farm, how to make composts, how to treat pests and diseases and how to work with the weather of different seasons. These are religious texts and are thousands of years old. Howard also read about King's studies and observations from his tour of China. In 1924 he was appointed as director of the Institute of Plant Industry in the State of Indore in central India and there he started to develop the 'Indore method of composting'. His book, *An Agricultural Testament*, was published in 1940.[6]

The core of Howard's ideas was that 'a fertile soil, that is, a soil teeming with healthy life in the shape of abundant microflora and microfauna, will bear healthy plants, and these, when consumed by animals and man, will confer health on animals and man'.[7]

Albert Howard regularly corresponded with Lady Eve Balfour in the UK and Jerome Rodale in the US. Balfour was the niece of the prime minister Arthur Balfour. She was one of the first women to study agriculture in the UK and took on her family farm in Haughley, Suffolk, during the farming depression of the 1920s and 1930s. In 1938 she corresponded with and met Howard and started to experiment with the new composting systems he was describing.

Lord Northbourne held a summer school and conference on his estate in Kent in 1939. He invited Ehrenfried Pfeiffer to come to this conference. Pfeiffer was the pioneer farmer developing the principles and practices of biodynamic farming from Steiner's lectures. Eve Balfour attended this conference on his Betteshanger Estate in Kent and subsequently wrote *The Living Soil*, which was published in 1943. She was part of the group that founded the Soil Association in the UK in 1946. Together they set up a magazine called *Mother Earth*, which is now called *Living Earth*. In the US Jerome Rodale set up the Rodale Institute in 1947 in Pennsylvania, and that became the founding centre of the organic movement in the US. He in turn helped to found the USDA Sustainable Agriculture and Education programme. The Rodale Institute coined the use of the word 'regenerative' in the context of agriculture, and regenerative agriculture is now arguably a form of sustainable farming in its own right.

It is notable that these seven people observed and learned sustainable farming methods from long-established traditional farming practices in China and India. They took this knowledge back to the US and the UK, eventually founding what became the modern organic movement. Through a modern lens we could see this as a form of cultural appropriation.

Masanobu Fukuoka in Japan developed what he called 'natural farming' methods with complex mixtures of perennial cropping, minimum tillage and grain crops. Although in practice his methods are very difficult to reproduce, his writing in *The One Straw*

Revolution prompted many people around the globe to reseed areas of impoverished soil.[8]

In 1972 Roland Chevriot invited the UK Soil Association, the US Rodale Institute, the French Organic Movement and the Swedish Biodynamic Movement to meet in Versailles to create IFOAM to link up their work, to formalise the registration of organic systems across borders, to share research and to build a worldwide movement.

In India Bhakar Save has reintroduced organic farming, which he calls 'natural farming', with complex polycultures of perennial and annual cropping, alongside a revival of Vedic farming practice, arguably the origin of organic farming.

Principles of organic food production systems

As late as 2005 a definition of 'organic agriculture' was agreed on by IFOAM:

'Organic Agriculture is a production system that sustains the health of soils, ecosystems and people. It relies on ecological processes, biodiversity and cycles adapted to local conditions, rather than the use of inputs with adverse effects. Organic Agriculture combines tradition, innovation and science to benefit the shared environment and promote fair relationships and a good quality of life for all involved.'[10]

IFOAM laid out the four principles of organic farming as:

- The Principle of Health: 'Organic farming should support and enhance the health of the soil, plants, animals and humans as one and indivisible.'

- The Principle of Ecology: 'Organic Agriculture should be based on living ecological systems and cycles, work with them, emulate them and help sustain them.'
- The Principle of Fairness: 'Organic Agriculture should build on relationships that ensure fairness with regard to the common environment and life opportunities.'
- The Principle of Care: 'Organic Agriculture should be managed in a precautionary and responsible manner to protect the health and well-being of current and future generations and the environment.'

IFOAM is the international body that oversees all of the registration bodies that inspect farms and food processors and give them permission to sell their food under the Organic Standards symbol. More specifically, the UK Soil Association laid out the principles of organic farming as follows:

- To produce food of high quality and in sufficient quantity by the use of processes that do not harm the environment, human health, plant health or animal health and welfare.
- To work within natural systems and cycles at all levels, from the soil to plants and animals.
- To maintain the long-term fertility and biological activity of soils.
- To treat livestock ethically, meeting their species-specific physiological and behavioural needs.
- To respect regional, environmental, climatic and geographic differences and the appropriate practices that have evolved in response to them.

Vedic farming and natural farming methods revival in India

Dr Prabhakar Rao lives near his farm just outside Bangalore. He practises natural farming, which is based on the ancient practice of Vedic farming. Vedic farming finds its roots in the Rig-Veda, a set of hymns and scripts written in Sanskrit and developed from the ancient culture in the Indus Valley 5,000 years ago. It is the traditional ecological knowledge of the region. Rao describes four key elements in Vedic farming:

- Continual observation by the farmer leads to farming practice that is continually adjusted to respond to the weather and pests and diseases.

- The creation of soils rich with microbes, which function well in the tropical and subtropical soils of India.

- Soil is always mulched.

- A huge diversity of plants are used to create complex polycultures.

When the soil is full of worms and microbes and mulched the Indian word used to describe them is 'satru'. It will be very resilient to hot dry conditions followed by monsoon rains. When India was under British rule in the early 20th century, the British insisted the Indian farmers should apply urea as a form of nitrate fertiliser. The Indian farmers said at the time that it would destroy the satru of the soil.

In practice, the creation of a soil full of satru uses certain preparations and mixes that are added to the soil and plants. The Ramen cow was the predominant cow in the Indus Valley and it has huge numbers of microbes in its gut. Many of these microbes have the ability to remove nutrients from the soil particles and make them available to plants. The Vedic system takes fresh cow dung and adds chickpea flour (a form of protein), jaggery (sugar) and soil. In a few days this mixture will have 'exploded' the number of microbes in the mixture and this is then poured on to the soil

in a dilute form. These microbes need to be protected from direct sunlight while receiving plenty of air and moisture. This is done by mulching. The earthworms aerate the soil and move the nutrients around. The name of this recipe translates into English as 'the elixir of life'. There are many crossovers here with the use of the biodynamic preparations.

There are lots of different mixes like this in the Vedic system. For pest control there are more solutions, called 'astras'. Leaves of trees that can repel insects, such as those of the neem, are collected and boiled five times in cow urine The urine acts as a solvent. This is diluted and sprayed on insects on plants. Boring pests are treated with a mix of garlic, tobacco and chillies boiled five times in cow urine, then filtered, diluted and sprayed on the plants. These treatments are very effective and do not instigate any resistance in the insect population.

Rao suggests that the Indore composting method developed by Albert Howard is not as effective as the composting systems prevalent in India. The heat generated in the compost heaps in the Indore method destroys the microbes in the compost, and so it will be less effective, especially in a tropical country like India. The organic matter will improve the soil but it will not enliven it with microbes.

Rao's farm is full of trees. Its rich soil is full of earthworms and fully mulched.

Virtually no tillage goes on. He suggests that this is what Indian farming looked like before the British arrived. He saves seeds from traditional Indian fruit and vegetables on the verge of extinction and sends them out to other farmers. He supplies 2 million farmers in India with seeds and together they cultivate 800,000 hectares of crops. His seed company is called Hariyalee Seeds.[9]

- To maximise the use of renewable resources and recycling.
- To design and manage organic systems that make the best use of natural resources and ecology to prevent the need for external inputs. Where this fails or where external inputs are required, the use of external inputs is limited to organic, natural or naturally derived substances.
- To limit the use of chemically synthesised inputs to situations where appropriate alternative management practices do not exist, or natural or organic inputs are not available, or where alternative inputs would contribute to unacceptable environmental impacts.
- To exclude the use of soluble mineral fertilisers.
- To foster biodiversity and protect sensitive habitats and landscape features.
- To minimise pollution and waste.
- To use preventive and precautionary measures and risk assessment when appropriate.
- To exclude the use of GMOs and products produced from or by GMOs with the exception of veterinary medicinal products.
- To sustainably use products from fisheries.[12]

Practices of organic food production systems

These principles have been developed into a set of Organic Standards that translate into a series of practices. In order to sell produce with the Organic symbol on it a farm or market garden first has to be registered and go through a period of conversion. Once these criteria have been met, normally over a two-year period, the crops and produce from that farm can be sold and marketed as organic and are generally sold at a premium price.

Those practices can be categorised under the following headings. They will vary according to soil type and region, but follow similar patterns. No chemically produced fertilisers, pesticides or herbicides are permitted and all the crops have to be grown in the soil.

Soil fertility

The 'rule of return' is the basis of creating soil fertility on an organic farm. Ideally, soil fertility is sustained on a farm by the use of composted animal manure, crop wastes and green manures. Green manures are plants grown specifically to put fertility and organic matter back into the soil. These are sometimes grazed with sheep, cows or chickens to add further fertility. Mixtures of legumes and grasses are commonly used – clovers, vetches, rye grass, buckwheat, to name a few. Extreme fertility imbalance occurs sometimes during the conversion of the farm or where the soil has a high or low pH. This causes soil mineral deficiency. In these circumstances some mineral supplements can be added. These supplements are normally a waste product of an ecological cycle. If a soil is alkaline, there can often be a magnesium deficiency, and in this scenario it is permissible to apply Epsom salts. Liming is provided by ground limestone, a naturally occurring lime.

Pest and disease control in plants

Crops are rotated to prevent the build-up of disease and to balance out fertility. This involves a minimum of four years for most vegetable crops. Often crop rotations are more complex.

The soil is sown with green manures and herbal leys and then grazed with animals such as cows, sheep or poultry to put fertility back into the systems and to break the pest or disease cycle. It can be as long as 10 years before a crop is repeated in the same place.

The aim is to farm in such a way that pests and diseases don't arise, but if they do arise an organic farm can use some interventions. It is permissible to use simple 'sprays' such as sulphur to counter fungal diseases, or to introduce biological predators to eat the pests.

Physical barriers such as horticultural fleece can be used to prevent rabbit damage or caterpillars eating brassica crops. The timing of sowings can also prevent pests; for instance, sowing carrots late in the season is a way to avoid carrot root fly.

Other pest control practices include choosing varieties of crops and breeds of animals that are suited to the place and climate. One example of this is the use of crops that are genetically resistant to diseases, such as apple varieties that are genetically resistant to scab, an apple disease. Another strategy is to use open-pollinated or genetically diverse breeds of crops, such as the population wheats, that slow down the movement of disease through the crop. This means pesticides are no longer necessary. Vegetable crops are preferably open pollinated, which gives them strong genetic diversity. Genetically modified or edited seeds are not permitted at all. Organic growers and farmers and livestock are obliged to source their seeds and stock from organic breeding suppliers. If they don't, they have to obtain a 'derogation' to explain the reason for their choice.

Pest and disease control in livestock

The choice of livestock breed, low stocking rates, and feeding the animals with good-quality organic food all reduce the amount of stress the animals suffer. This in turn reduces pest and disease incidence. With sheep and cows the aim is to grass-feed them as much as possible. In the winter this is supplemented with organically registered hay or silage grown on the farm or bought in. Giving livestock the opportunity to graze on a diversity of grass, herbs and trees improves their health and diet and allows them to self-medicate for worms or other imbalances as they would in the wild or before domestication. Animals have the innate ability, given the opportunity, to choose plants to eat that will cure diseases and pest problems. By the correct choice of herbal leys and agroforestry trees the need for animals to be dewormed can be reduced to zero.

Chickens and pigs are kept outside on pasture and ideally fed with food from the farm, although most organic farms buy in feed for these animals from organic suppliers. High animal welfare is required: chickens are not allowed to have their beaks cut, and pigs are not kept in farrow cages. These practices are unnecessary once the stocking rates are reduced and the animals are allowed to be outdoors, since they will be less stressed. The animals are treated with homeopathic medicines.

If an extreme outbreak of a pest or disease happens then antibiotics can be used, with a derogation, and the produce is removed from sale for that period.

Weed control

Weeds are controlled by mechanical weeding with tractors or by hand, depending upon the scale. The use of green manures can reduce the presence of weeds in a field. Cultivation methods such as 'stale seedbeds' reduce the weed seed burden in the soil; this is where the seedbed is prepared, the first flush of weeds are allowed to germinate and then these are shallowly cultivated or burned off by flame weeding. Perennial weeds such as couch grass are controlled by stale seedbeds over the growing season. Mulches are used wherever possible to shade out weed growth.

Closed loop

Organic farming is sometimes equated with agroecology but there is one big difference: as a practice, organic farming does not insist on the creation of closed loop systems. Although many organic farmers do work towards creating a closed loop system, it is not a requirement for the Organic symbol.

Miguel Altereri, one of the main academic thinkers in agroecology, points out that organic farming started out looking very similar to agroecology. As the organic farming movement has grown and developed it has in some places started to resemble the industrial model of farming. Some large organic farms use an input–output model rather than a circular farming model. They can buy in soil nutrients, but from biological sources such as dried chicken manure pellets, and buy in more benign equivalents of pesticides such as sulphur or pyrethrum, and sell their products to a distant supermarket, even fly them across the world wrapped in plastic. Organic produce tends to be sold to high-income customers, and if it is grown at all in the global south it is an export crop for wealthy customers in wealthy countries. Altereri suggests that the organic movement could have avoided this by limiting the size of farms, allowing more flexibility in standards in different regions, and including social standards for the welfare of workers (which the UK standards do).[13] Hindsight often offers more clarity than the pioneers of organic systems had in the 1960s and 1970s. The question of how sustainable organic farming actually is brings debate about whether it is better to buy organic food or local food. The real point, perhaps, is that all these systems need to work together to create an overall food system that is more sustainable and healthy than the industrial model.

Do organic food production systems work?

There are many studies that show that organic food systems work well. They grow good-quality food with high nutritional value. They have high welfare provisions for livestock. Food sold as organic can command a premium price. The yields of organic foods are generally 20 per cent lower than those in industrial farming systems for some but not all crops. This works out financially for the farmers because the inputs are lower and the product generates a premium. Professor Jules Pretty, of the University of Essex, demonstrated that organic farming provides £209 of environmental benefits per year for every hectare in production in the UK.[14]

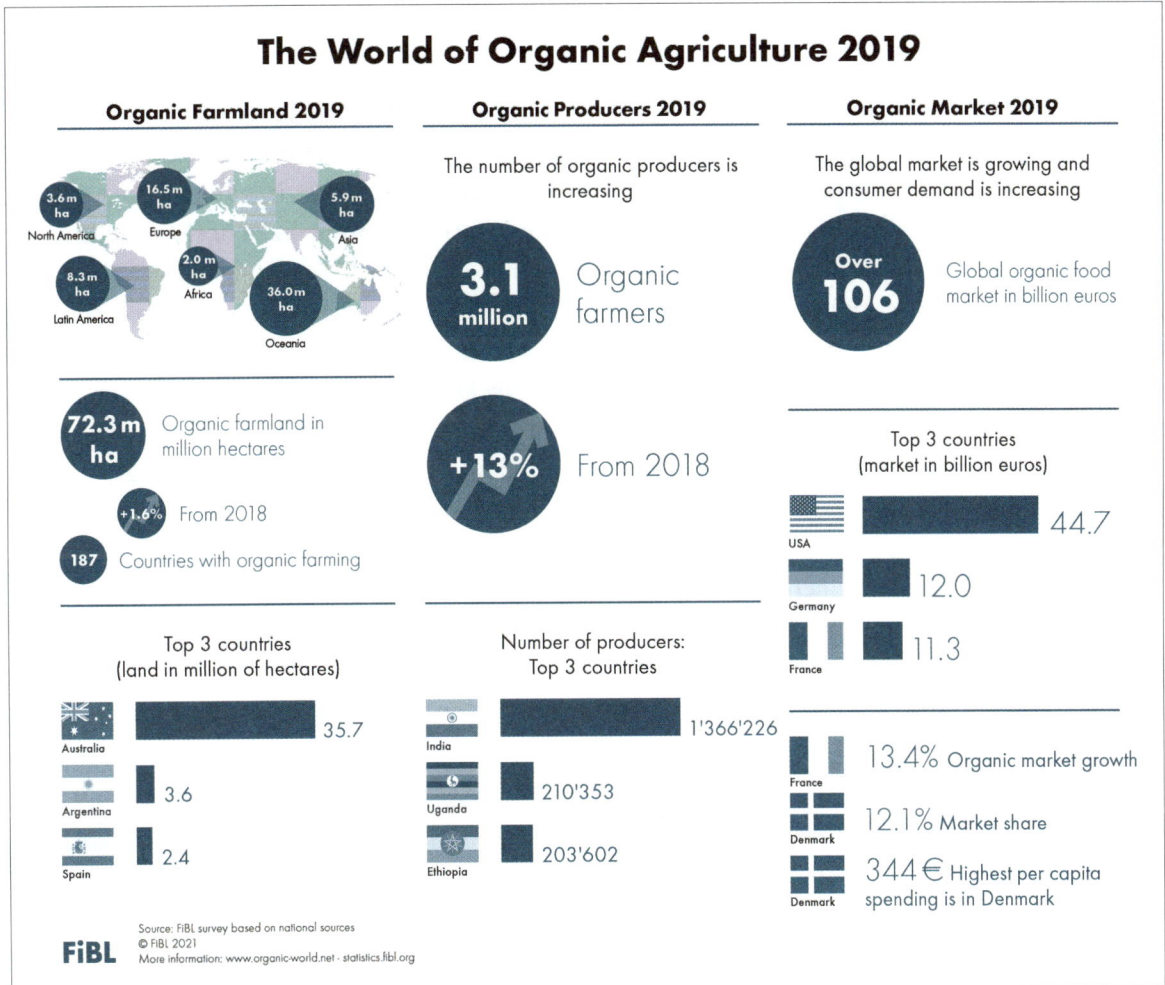

Figure 4.1: World organic agriculture infographic from FiBL.

An 'organic agriculture' report from the FAO suggests that farms practising industrial farming for long periods of time show a decline in yields as soils become exhausted.

However, in 'green revolution' areas of the world that switch from industrial to organic or agroecological systems, yields often remain the same. Those 'undeveloped' regions that skip the industrial farming stage and move straight on to organic or agroecology systems experience increases in yields of crops of up to 130 per cent.[15]

Many studies and reviews of studies have shown that organic farming sequesters more carbon, up to 25 per cent more than its industrial counterparts.[16] Organic farms support more biodiversity.[17] The food they produce is more healthy, with higher mineral levels, more secondary metabolites and no pesticides.[18]

What does an organic food production system look like?

Organic farms range from tiny urban 'patchwork' farms in city centres, to mixed mid-sized family farms in rural areas, peri-urban market gardens supplying local customers, restaurants and shops, and large farms that are thousands of hectares in size and supply supermarkets across the world.

Across Europe almost eight per cent of farmland is certified organic, Liechtenstein having the highest proportion at 38 per cent. Across the globe 71.5 million hectares are farmed organically, that is, 1.5 per cent of all farmland. Organic land has increased year on year since 1999 by more than 500 per cent. The global market for organic food is worth €95 billion.

Most of the organic food is eaten in the US and Europe. Countries in Africa and Asia often supply export crops into the global north. As middle classes develop across the world the market demand for organic food increases. Australia has the largest total area of land in organic production. India has the highest number of organic farmers. The Indian state of Sikkim has converted 100 per cent to organic farming. The US is the largest consumer of organic food in the world. IFOAM and the Swiss Federal Institute of Biological Agriculture (FiBL) publish an annual interactive map and set of global statistics on their website (Figure 4.1).[19]

Case study: Tamarisk Farm

Tamarisk Farm is found on the hills sloping down to the coast in Dorset, in the southwest of England (Figure 4.2). It is an organic mixed family farm operating on 72 hectares, with a further 161 hectares rented from the National Trust. The main enterprises are a 25-cow beef suckler herd of Red Devon cattle and 180 ewes, mainly Dorset Down, Shetland and Jacob. There are 36 hectares in arable rotation, growing wheat, rye, barley, oats and drying peas.

The land is poor quality, but has a diverse range of wildlife habitats as a result. The blackthorn scrub is kept at bay by running cattle and sheep extensively over Tamarisk's pastures, and a wide range of wild flowers, insects, birds and reptiles are able to live alongside the livestock. The farm is home to nine species of orchid, grass vetchling and dyer's greenweed. Resident and migratory birds, including kestrels, barn owls, hobby and corn buntings, as well as dormice and great crested newts are all found on the farm. The livestock breeds have been chosen for their ability to thrive on poor-quality grazing and are fed exclusively on fresh grass, hay and a very small bit of homegrown grain for the ewes prior to lambing. The farm is currently supported by the UK government's higher level of subsidy intended to protect wildlife, called the Higher Level Stewardship Scheme.

To remain viable, the farm has developed a range of income streams, centred around producing high-quality organic meat, cereals and wool for local markets. A small farm shop sells meat, home-milled stoneground wheat

With kind permission of Walter Lewis

Figure 4.2:
Tamarisk Farm

and rye flour, peas, free range eggs from hens fed on home-grown cereal, and a wide range of naturally coloured wools. Other produce is sold into local supply chains, such as cleaned grains to local water mills or meat to local butchers' shops, restaurants and abattoirs. Many of the cattle are sold as store cattle when one year old to the Organic Livestock Marketing Company. Tamarisk Farm sees public engagement as an important part of its role. Although the farm no longer receives support for visits other than

ones by school age children, educational farm events are offered to people of all ages.

Alongside the farm's annual open day, a series of visits focused on things such as lambing, the arable and milling enterprises, wild flower identification and the ways wool can be used give visitors an opportunity to learn how food production and nature conservation practices can occur simultaneously on the same piece of land.[20]

Case study:
Sarah Green Organics

Sarah Green is the third generation to farm at Mark Farm, Tillingham, Essex. The farm is situated on the east coast of the Dengie Peninsula, between the North Sea and the rivers Blackwater and Crouch. It has 69 hectares of land on sandy loam to sandy clay loam.

When Green finished her training at Writtle College she returned to the family farm and decided to start an organic vegetable business on 10 hectares registered with the Soil Association. This has now grown to 25 hectares of mostly field-scale vegetables all year round (Figure 4.3). Green specialises in winter vegetables, since not many other growers were doing this. This has given her business an all year round access to markets in London and more locally in Essex. She employs 14 people excluding her husband and parents, who also work on the farm.

Soil fertility is managed with green manures and leys, and grazing by a small flock of Poll Dorset sheep. Lamb is sold locally by word of mouth to friends and other customers. Compost is also made and spread before ploughing in the spring. The pests and diseases are controlled by good crop rotation. The population of natural predators is built up by the use of bait plants and the provision of overwintering habitat. Crop covers prevent infestation by pests such as the cabbage white on cabbages; they prevent the adult butterfly from laying eggs on the cabbage. Biological controls are used in the polytunnels when required, such as bought-in *Encarsia formosa* that eat whitefly, a pest on crop plants. The vegetables are sold weekly via a vegetable box scheme and delivered up to 55 km from

the farm, with collection points at the furthest locations.

Growing Communities have developed the Better Food Shed, with a distribution hub on the outskirts of London. Growing Communities themselves pack 1,000 vegetable bag deliveries in London, mostly in the Hackney area. They have helped other organisations to set up similar box schemes in other boroughs of the city; they source their vegetables from this same distribution hub. Sarah delivers to the Better Food Shed weekly and to other wholesale customers midweek.[21]

Figure 4.3. Sarah Green's organic farm.

Training in organic farming methods

There is very little state-funded specific training in organic farming in the world. There are some degrees across Europe, although these are often labelled as 'agroecology' or 'sustainable farming', of which organic farming is a component.

Organic associations such as the Soil Association in the UK run seminars and webinars for farmers wishing to convert to organic systems, and run farmer-to-farmer training sessions.

Where is the organic food production movement going?

IFOAM published its Phase Three report about the organic movement in 2016. In it they reflect that the organic movement has developed in three phases. Phase One was the pioneering phase, when the concept of organic farming was developed 'by those who saw the connections between how we live, eat, and farm, our health and the health of the planet'.

Phase Two was the creation of the movement, the standards, the principles and practices and the codification that have established organic farming in 87 countries across the world. The organic market was estimated to be worth €95 billion per year in 2018.[22]

However, the organic movement recognises that after a little less than 100 years it has only managed to occupy less than 1.5 per cent of the world's farmland. Organic farming can exclude those who need it the most, such as women farmers in the global south. Building bridges with other sustainable practices such as agroecology has also been problematic because of the certification requirement of organic farming. These considerations inform IFOAM's aspirations for Phase Three of the organic movement:

Agriculture should be a force for good, providing solutions to global issues of hunger, inequity, energy consumption, pollution, climate change, loss of biodiversity and depletion of natural resources. The positive, multifaceted environmental, social and economic benefits of a truly sustainable agriculture can contribute solutions to most of the world's major problems. If mainstream agriculture adopted truly sustainable practices, the need for certified organic agriculture would cease to exist. Until now, though, organic has not been included – or inclusive – enough to contribute these solutions on a global scale. The Organic Phase Three concept seeks to change this, by positioning organic as a modern, innovative system that has positive impacts on the above-mentioned issues.[23] ∎

In a nutshell

- Organic farming is based on the rule of return but is not obliged by the codified standards to aim to be a closed loop system.

- Waste products and manure are composted and returned to the soil to build fertility.

- Pests are managed by biological and cultural methods, that is, the use of natural predators, crop covers and genetic resistance.

- Livestock are kept in high-welfare, low-stocking conditions.

- Food is codified, that is, traded with a registered trademark, and can generate premium prices.

- Organic food is sometimes sold via local trading systems such as CSA or farmers' markets.

- The food is sometimes shipped or flown across the world and sold in supermarkets.

Further reading, links and films

The Riverford Farm website hosts 'Wicked Leeks' an online organic farming news feed, with films, articles and links relating to organic and sustainable food.

J. Moyer, A. Smith, Y. Rui, J. Hayden, 'Regenerative Agriculture and the Soil Carbon Solution', Rodale Institute, 2020.

A short film about Bhaskar Save is *A Gandhian Way of Farming*: https://www.permaculturenews.org/2014/02/26/bhaskar-save-gandhi-natural-farming

1   E. Balfour, *The Living Soil*, Faber & Faber, London, 1943.

2   https://www.ifoambio

3   P. Conford, *The Origins of the Organic Movement*, Floris Books, Edinburgh, 2001.

4   F.H. King, *Farmers of Forty Centuries or Permanent Agriculture in China, Korea and Japan*, Jonathan Cape, London, 1927.

5   Conford, *The Origins of the Organic Movement*.

6   A. Howard, *An Agricultural Testament*, Oxford University Press, London, 1940.

7   L. Howard, *Albert Howard in India*, Rodale Press, Emmaus, PA, 1954, p. 162.

8   M. Fukuoka, *The One Straw Revolution*, New York Classics, New York, 2009.

9   P. Rao, 'Natural Farming: It's Been Handed Down to Us', Bangalore International Centre, 21 October 2019.

10 https://www.ifoambio

11 Ibid.

12 https://www.soilassociation.org

13 M. Altieri and C. Nicholls, *Agroecology and the Search for a Truly Sustainable Agriculture*, UNEP and PNUMA, Mexico City, 2005.

14 J. Pretty, *Agri-Culture*, Earthspan, London, 2002.

15 http://www.fao.org/organicag/oa-faq/oa-faq7/en/

16 Soil Association, 'Soil Carbon and Organic Farming', 2009.

17 Organic Research Centre (ORC), 'Biodiversity Benefits of Organic Farming: New Research Confirms More Diversity on Organic Farms', 2010.

18 Soil Association, 'Organic Farming, Food Quality and Human Health', 2001.

19 FiBL, 'The World of Organic Agriculture Statistics and Emerging Trends', 2020.

20 tamariskfarm.co.uk

21 https://www.sarahgreensorganics.co.uk

22 https://www.organicseurope.bio

23 IFOAM, 'Organic 0.3. For Truly Sustainable Farming and Consumption', 2016.

Chapter 5

Permaculture food production systems

We already know how to build, maintain, and inhabit sustainable systems. Every essential problem is solved, but in the everyday life of people it is hardly apparent.[1]

Bill Mollison

Introduction

Bill Mollison was one of the founding fathers of the permaculture movement and he wrote those words in 1988; today they are truer than ever. We know how to live sustainably on the planet but for many complex reasons we choose not to. The permaculture movement has collated sustainable systems from around the world in a coherent design methodology. A series of permaculture sites small and large all over the world are demonstrating good sustainable food production as a form of permaculture practice.

Permaculture in its simplest form is a powerful methodology that can be used to design farms. It allows farmers to reassess and design their farms so they are more sustainable and economically viable, and at the same time socially connected. It is a practice, a movement, a philosophy and at times it is political. This chapter will focus on the principles and the practice of the design methodology, the form this takes on a farm, and where you can visit a permaculture-designed farm in real life or via films or books. The chapter will also examine the scientific proof of how effective permaculture food systems are and signpost you to the other aspects of the permaculture movement.

What is permaculture?

The name 'permaculture' arose from the fusion of the two words 'permanent' and 'agriculture'.

Its purpose is the conscious design and maintenance of agriculturally productive ecosystems that have the diversity, stability and resilience of natural ecosystems. [Permaculture] is the harmonious integration of landscape and people providing their food, energy, shelter, and other material and non-material needs in a sustainable way.[2]

It has now been applied in the creation of farms, smallholdings and gardens in over 140 countries.

As with many of the other farming systems outlined in this book, those practising permaculture have acknowledged previously existing indigenous systems, observed the essence of them, woven them together and translated them into a set of principles and practices that can be replicated as a model for recreating sustainable systems all over the world.

Permaculture was originally land based but has now grown to include non-land-based systems as well, such as websites, businesses, healthcare and communities. All these systems can be created using the design methodology, principles and practices.

The design methodology follows the process of Survey, Analysis, Design, Implementation, Maintenance, Evaluation and Tweak, summarised in the acronym 'SADIMET'. There are different iterations of this process with different acronyms, but they are all essentially the same process.

The concept is that, through careful observation, analysis and design, a productive closed loop system can be designed, in which most of what is needed on any given site by the people, animals and plants that live there is grown on that site. All the waste outputs created are used on site as an input back into the system, and the surplus produce is sold or shared.

Pollution is regarded as wasting resources that need to be fed back into the system to create an abundant closed-loop farm, or not produced in the first place. These systems are designed to have positive feedback loops, so that they will evolve and develop in a 'spiral of abundance'. A great deal of emphasis is placed on careful planning and implementation with the aim that a permaculture-designed system will require much less maintenance work than an industrial or even an organic farming system. Mollison says that these systems are much kinder to people, and that with the resulting spare time we can 'sit on our veranda and play the guitar'. This is a very beguiling idea for most hard-working farmers and food producers. Underlying these systems is the premise that, once created, they are highly productive, and we can use this fact to improve our quality of life.

So how does the design process work? Mollison said, 'Permaculture design is a system of assembling conceptual, material and strategic components in a pattern which functions to benefit life in all its forms.'[3] The underlying concept behind this design methodology is surprisingly simple. A farm or garden can be viewed as a series of 'zones', from the one that the farmer or gardener spends most time in, Zone 0, through to Zone 5, which is visited least frequently (see Figure 5.3, page 80). Zone 00 is the farmer themself.[4]

The concept is to arrange the things, or 'elements', that we want on the farm (the elements of the design) into these zones according to how much human energy they require or the frequency of visits. The areas of the farm that need the most energy or frequent visits are placed closer to the centre, Zone 0. Those that require fewer visits are placed further from the centre. This means that the most efficient use is made of the human energy on the farm.

The term 'sectors' refer to the natural energy flowing into the site from outside (see Figure 5.4, page 80). The aim is to use as much of this energy as possible before it leaves the site, within each

Joanna Macy and the Great Turning

Macy suggests that three important types of work are required to bring about lasting change from the industrial growth paradigm that we currently live in to the low-impact sustainable paradigm that we need to move towards. There need to be those people who halt the destruction and degradation of the environment, those who document and protest against the powers and forces that are destroying the planet, and creating climate change, and those who provide analysis of the structural causes and provide alternatives to the current industrial model.[6]

zone, or to deflect the energy if it is harmful. The sectors that are observed and mapped are the wind, sunshine, frosts, views, the flow of people, roads and paths and the flow of water. The sectors are unique for each site, making each permaculture design site specific. Ideally, each site will be observed and recorded for a year before the design takes place. This might result, for example, in harnessing the incoming wind to generate energy via a wind turbine, or planting windbreaks to shelter crops from cold or damaging winds. Or both can be done in some cases.

Each element of the design is placed in the right zone and in the right sector of the site to make the most of the human energy and the energy coming into the site from outside.

The 'functions' of the farm are clearly identified and these are specific to the farmer. The list often includes functions such as 'produce food', 'support biodiversity', 'create an income', and it may include providing 'educational facilities'.

The 'elements' are the things that are to be included on the farm. Elements include things like chickens, bees, orchards, fire pit, and barn. Mollison said, 'the philosophy behind permaculture is one of working with rather than against nature, or protracted observation rather than protracted and thoughtless action; of looking at systems in all their functions, rather than asking only one yield of them; and of allowing systems to demonstrate their own evolution'.[5]

To create a resilient system each function of the farm needs to be supported by more than one element and each element should support more than one function.

Where did it start?

Permaculture originated in Tasmania in Australia in the early 1970s. Its two founders were Mollison and David Holmgren. Mollison was born in 1928 in Tasmania. He left school young, working in his family bakery, and then moved on to working for the Australian Commonwealth Scientific and Industrial Research Organisation, in the wildlife survey section deep in the Tasmanian forests. He was very distressed at the destruction of old growth forests that he witnessed. He studied for a degree in biogeography late in life and then became a lecturer in, among other things, environmental psychology at the University of Tasmania. A passionate and charismatic man, he left his secure lecturing post and, with Holmgren, created permaculture. What they came up with was a systems rethink to provide an alternative to the large-scale industrial farming that was and still is eating up vast amounts of land and old growth forests across the world. They wove together principles of landscape design with agriculture and ecology and created something unique.

From Joanna Macy's work on 'the Great Turning' we can understand that a common journey for environmental activists is, initially, a path of protest, followed by one of creating positive solutions that address the underlying issues. Protests to stop environmental destruction are vital but people need an alternative path to travel down to build a new sustainable future. Mollison and Holmgren, too, started out as protesters. Their switch to creating an alternative path has given many millions of people worldwide the ability to create thousands of sustainable systems.

Mollison wrote a series of books, notably *Permaculture: A Designer's Manual* in 1988. He then took to the road to train people worldwide via the permaculture design course. He set up Tagari Farm and Permaculture Research Institute in Australia, which has now moved to Zaytuna Farm, also in Australia. Holmgren trialled the systems on a small 0.8-hectare site called Melliodora in Hepburn Spring, Victoria, which is still functioning. Once he was confident the systems worked, he also wrote a book called *Permaculture: Principles and Pathways Beyond Sustainability*, set up a design practice and joined the training circuit.[7]

Principles of permaculture food production systems

The permaculture movement has created a clear set of ethics and principles that are used in the decision-making process when designing a farm or garden.

Permaculture ethics are simple: care of the earth; provision for all life systems to continue and multiply; care of people; provision for people to access those resources necessary for their existence; setting limits of consumption. By governing our own needs, we can set resources aside to further the above principles.[8]

The **earth care** ethic suggests that decision-making when designing a farm should serve to benefit the earth. That implies that the farm will actively mitigate climate change through

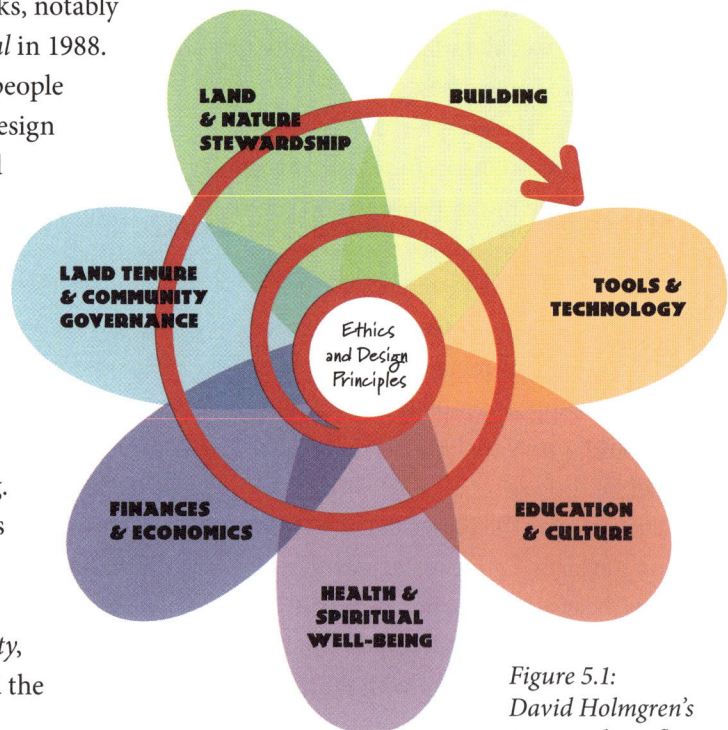

With kind permission from the Permaculture Association.

Figure 5.1:
David Holmgren's permaculture flower.

maximising carbon sequestration and supporting biodiversity. This can range from the way in which the farm is designed to simple choices of materials when implementing a design.

The ethic of **people care** suggests that a permaculture farm should be socially just. This implies that staff will be paid living wages and that working conditions will be fair. It may encourage the farm to provide access for educational or recreational visits.

The **fair share** ethic refers to people only taking as much as they need from a system and leaving the rest for other people or for wildlife. The underlying belief is that there are enough resources to go around in the world, but that the global north is overconsuming resources at the expense of the global south or the future generations. For a farm this ethic may imply that the business structure should be set up as not for profit, in order to keep prices of food down to customers, or that gluts will be donated to food banks to reduce food waste.

The permaculture principles divide into two groups: those about attitude and those based on ecology. These principles are implemented in farms that are designed on the basis of permaculture systems. Two main sets of principles are in use: Holmgren's and Mollison's.

These broadly align but for ease I have used Mollison's principles.[10]

Firstly, there are principles of attitude

1. **Work with nature rather than against it**
   Nature is powerful and we understand from the study of ecology that nature follows a path of ecological succession. Soil will try to cover itself with plants, grasses and annual plants, which will lead in turn to perennial plants and eventually trees. The final stage of ecological succession is called 'climax vegetation'. Farmers and food producers try to hold back this process, by weeding, trimming, cutting, mowing, pruning, which is hard work requiring a lot of energy. By observing this natural process we can create food-producing areas that travel in the same direction as nature and so require less energy. If soil is mulched it does not need weeding and requires less water. If we create a forest garden like an edible form of 'climax vegetation' it requires less maintenance. If arable farms use deep-rooting complex polycultures of grains they need less weeding.

2. **The problem is the solution**
   This principle is about a creative approach to problem-solving. When designing a farm or a garden the normal approach would be to say, 'I want to grow apples in that field.' If the field is wet and slightly acidic then the field will be drained and the farmer will add lime to change the pH. These are invasive and expensive solutions to the problem. The permaculture farmer will ask, 'What can I grow in wet acidic soil?' The answer may be willow coppice for basket-making or biomass burning, or ramial wood chips for increased soil fertility, or fruit that like wet acidic soils, for example blueberries and bilberries. (Ramial wood chips are thin branches of trees such as willow that are chipped and poured on the soil. These wood chips are high in plant fertility and organic matter.)

3. **Make the least change for the greatest effect**

When there's a problem on a farm without an obvious solution, there is a tendency to use large fast solutions that are quite expensive and intrusive. The principle here is instead to observe the problem, try out small simple solutions, observe the feedback, and the solution will become clear.

Let me illustrate this with an example from Huxhams Cross Farm. We had a very wet field, with poor soil structure. When it rained, the water ran off the surface, taking a lot of top soil with it; it did not percolate into the soil. The standard agricultural process would be to mole drain the ground with large, expensive and heavy equipment or just leave it fallow and unproductive, as this land had been left for the last 10 years, sprayed with glyphosate by the previous farmer to keep the weeds down. We cultivated the field and sowed it with deep-rooting green manures, put on biodynamic preparations to add biological activity, and keyline ploughed the field to let the air in and let the water out. The field slowly came back to life.

We planted it with alders and they thrived. We planted 50 fruit trees and they didn't die, so we planted a further 150 the following winter. We carried on observing the really wet patches of the field, which we intend to become rainwater-harvesting 'gardens', planted with wet-loving species such as willow, able to hold on to the winter water well into the summer.

4. **Everything gardens**

Permaculture asks what can be 'stacked' into a system to make it more productive? An orchard can produce fruit but it is also a great place to keep chickens and bees, and hence increase an income stream potentially threefold. The trees, chickens and bees provide many benefits to each other. The chickens will do some work for the food producer by eating pests and keeping the grass short so it does not need to be mown. The bees pollinate the fruit trees.

When the idea of 'Place elements to maximise the beneficial relationships between them' is applied, the food producer's workload is lessened. If the chicken house in the orchard has gutters and harvests rainwater with a self-filling water trough then there will be even less work, and the farmers can go and play the metaphorical guitar on the veranda.

5. **The sum of the yields is limited by your imagination**

Industrial food systems think of yield in one way only, as a yield of a crop or animal that can be sold. This term can be broadened to imagine a 'yield' of biodiversity or carbon sequestration, or the yield of happy children on a farm. Farms can be designed to have many multiple yields and benefits. This is called 'natural capital' by environmental economists. By counting the natural capital, social capital and human capital as well as the financial capital in a system, accountants can assess the whole value of a system, not just one part of it.

6. **Cooperation and not competition**

In modern Western society the prevailing thought is that competition is vital in order to survive and thrive. Charles Darwin coined the phrase 'survival of the fittest'. This has been the prevailing concept of how ecosystems work. It is now understood that collaboration and cooperation are far more prevalent between species than Darwin thought. In *The Hidden Life of Trees* Peter Wohlleben describes how trees have partnered up with fungi on their root systems, enabling the trees to absorb more water and nutrients from the soil and send messages and share resources with neighbouring trees. The fungi get fed with photosynthetic sugars that the trees produce and leak out of their roots. The fungi actually grow into the root hairs of the trees and form dense networks in the forest soil. Permacultural food producers can also learn to work together in similar ways with other food producers and customers to create better and stronger food systems.[11]

7. **Use biological resources where possible**

Use materials that are local and natural when possible. For example, if you are creating a fence it is preferable and often cheaper to use local materials, for example by making a woven willow fence or planting a hedge. The materials for building a house can literally be grown on a farm if the house is built from straw bales, timber framed, and plastered with mud and lime plaster and natural pigments. This gives the place a natural, vernacular look.

8. **Pollution is a wasted resource**

The waste stream from a farm can be fed back into the system. Flushing the toilet? Human manure can be fed into a reed bed full of nutrients and biodiversity. Throwing out paper? This can be shredded and used for chicken bedding. Creating closed loops systems by cycling nutrients, water and energy reduces resource use.

9. **Use and maximise diversity**

Increasing the diversity at every level of a food system increases the resilience of the system to pest, disease and extreme weather events. This may be on a cropping level – why grow one variety of carrot when you can grow five? Or on a genetic level, such as planting open-pollinated vegetable seeds. It could be on a species level, by creating polycultures, rather than monocultures; growing fruit, vegetables aznd cereals rather than just one range of products. Poly-incomes also increase resilience of income. A farm may produce crops and also offer educational visits to bring in extra income.

10. **Stacking**

Crops can be stacked in a food-producing space to increase yields. We often think of a field or garden in two dimensions, that is, as a flat space, but by adding in a vertical layer an extra yield can be obtained. This is seen in agroforestry practice (see Chapter 6). On a smaller scale, walls can be used to grow fruit trees, glasshouses can have trees and climbers in them, and fields of wheat can have trees in them.

11. Maximise the edges of the system

The greatest diversity of an ecosystem is found on the edge where two ecosystems interact. The length of the 'edge' between one ecosystem and another can be increased by putting in a wiggle or lots of thin strips, such as strips of trees, agroforestry rows, or hedgerow through a farm. If this principle is applied to the permaculture food system, agroforestry rows can be introduced into a wheat field. This will increase the length of edge of the rows of trees and wild grasses through the wheat crop. The agroforestry rows and wild grasses are home to many species of beetles and predatory insects such as ladybirds and hoverflies. These venture into the wheat crop and eat pests. They behave as 'functional biodiversity' in the system.

12. Efficient energy planning use zones, sectors and slopes

Farms and gardens naturally catch a lot of sunshine, water and wind. Ideal for small-scale power generation and rainwater harvesting. Buildings can be fitted with photovoltaic (PV) panels and wind turbines. Rainwater can be harvested at the top of the system and be gravity fed as irrigation water in the summer months. Elements are located relatively to each other to maximise the benefits. Putting a rainwater-harvesting system on the chicken houses can minimise the need to carry water to them. Putting the compost heap in the chicken area does away with the need to turn the compost heaps, because the chickens do the work of turning the heap when they scratch and forage for food.

13. Plan for the evolution of the system

A bare piece of earth being left entirely to itself would over many years regenerate into a climax woodland. Animals would move in and browse the wood. It would become like the 'wild wood' that is now found at Knepp Castle, as beautifully described by Isabella Tree in *Wilding*.[12] Food producers spend a lot of time and energy preventing natural regeneration from happening. We weed, turn the soil over and cut back hedges to stop nature doing what it wants to do. The evolution of a piece of productive land can be designed from the perspective that the farmer too is a part of the ecosystem and also needs to eat. A farm system can be designed that allows nature to do what it wants while still allowing the farmer to produce food. In a natural ecosystem there is never a bare patch of earth; it will soon be covered with plants that food producers often call 'weeds'. It is better for the soil to be covered with plants and not bare to the sun, rain and wind. A permacultural food producer aims to cover the soil with mulch, such as wood chips, compost or straw, or else will switch to more perennial crops that keep the soil covered. Green manures can be used between annual crops or as a follow-on crop, providing fertility and bee fodder and outcompeting the weeds. The weeds can also be eaten, since many of them are valuable forage wild crops. The farmer is thus able to spend less time weeding and can spend more time playing that guitar on the veranda.

A thought experiment on designing a farm or food system goes like this: If the farmer became very busy on one part of the farm this year, would the rest be lost to weeds? If the farmer walked out of the farm today, closed the gate and did nothing, what would happen? If the answer is that the soil would increase its microbial activity and organic content and the trees would grow, then the design has worked. If we find in our thought experiment that the farm system we have created follows a positive feedback loop, or 'spiral of abundance', then we know that we have created a regenerative farm system.

14. Create patterns; design from pattern to detail

When designing a large piece of land it is useful to start with the overview and then focus in on each aspect of the design with increasing detail in each iteration of the design cycle. For instance, in the pattern design of a farm a field may be marked 'orchard' and 'chickens'. The detailed designing will be done by going around the design cycle again, encompassing row orientation, spacing of trees, choice of varieties, cross-pollination, choice of sward, position of beehives, and windbreaks. To enable the 'chickens' to be in there, the rows will need to be wide enough for the chicken house, the sward suitable for chicken fodder, the sequential harvest able to let the chickens move through clearing the pests without getting in the way of the harvest. This approach enables the designer to create the broad brushstrokes of the design and enables the concept of closed loops to be built in early on in the process. If

these elements are made as multifunctional as possible the closed loop becomes a positive feedback loop, building fertility, biodiversity and water-holding capacity. For example: a farm needs fertility, and this can be produced on a farm by growing green manures. If clover is chosen as a green manure, it will also provide bee fodder, and chicken food, which will in turn add more fertility to the ground. It will also add more organic matter, increasing the water-holding capacity.

15. Build resilience into the system

Food production systems can be made more resilient by making sure that, as Mollison said, 'Every element performs multiple functions' and 'Every function is supported by multiple elements'. For example, in an irrigation system on a farm, the function of 'water for crops' can be served by rainwater harvesting, springs, boreholes or mains water; water loss can be reduced by mulching and increasing the organic content of the soil. These measures make the farm resilient if there is a drought. The element of a rainwater-harvesting system can support many functions, such as crop irrigation and encouraging biodiversity.[13]

Practices of a permaculture food production system

How does the design process work?

Permaculture is a holistic design process and as such in practice it needs to be a whole system if it is to function properly. You may consider a 'permaculture area' to try it out as part of a larger land project, but this misses the point. There is often confusion about growing techniques commonly used by permaculturists such as the forest garden and hugle-culture beds for growing vegetables. These are not in themselves 'permaculture'; they are only parts of a whole system and are practices that are sometimes used but not always. (Hugle-culture beds are deep-rooting raised beds that store lots of fertility and water.)

The 'practice' of permaculture is in essence that of designing a farm, or garden, that will demonstrate the permaculture principles. It will

Table 5.1: Site survey tools

Process	Description	Tools / Methods
Protracted observation	The farmer makes observations over a long period of time to get to know the site in different seasons and to build a personal relationship with it.	Photos Historical research Phenological diary
Slopes	Measure and record slopes	Water level surveying equipment Topographic map
Orientation of the site	Determine the aspect of the site in relation to sun and shade at the winter and summer solstice and the equinoxes.	Compass Topographic map Observation of the sun's movement
Soil	Understand the qualities of the soil, the opportunities it presents, its limitations and variations across site.	Soil ball test and jar test . Local knowledge . Professional lab testing of soil texture, organic matter, microbial activity and phosphate and potassium content.
Microclimates	Map microclimates on site, coldest and warmest temperatures in particular.	Observations and phenological diary Local weather records
People desire lines	Observe where people and animals walk on a site to create natural paths.	Observations and worn paths
The context of the farm	What is everyone doing around you? Building materials, crops, methods, techniques.	Asking questions of those around you Looking at historical maps
Vulnerable users	Who is using the site? Elders? Children? Notice their needs.	Client interview – the farmer and people who will be working on the farm
Mapping resources	What resources on site are to be kept or are found already.	Making a note of plants, animals, structures, tools, events
Mapping boundaries	Explore the boundaries of the site and make a base map.	Tape measure, pacing, drawing board, scale ruler, string, compass, paper Online tools: MAGIC maps, Google Earth

create a farm, that is, as Mollison defined it, 'an agriculturally productive ecosystem that has the diversity, stability and resilience of natural ecosystems.'[14]

Not all farmers are skilled enough in permaculture to design their own farms, and not all permaculture designers are skilled farmers or gardeners. Sometimes permaculture designers work with farmers and food producers to design a farm. Sometimes a farm will be designed by the farmer or grower, along with their community or farm team. Either way, the design process follows the steps below.

1. The survey.
Both the farm or site and the client are surveyed. If you are designing your own farm or garden the client is yourself, and your farm team or family. Alternatively, a professional designer may be working with a farmer or a community group. The survey process gathers unique information from both the site and the farmer using the methods presented in Table 5.1.

The next stage the farmer or designer will carry out is the client survey, even if the client is themself. Many permaculture farms will have other interested parties as well and these need to be included in the consultation process. The aim is to create a list of the functions and elements needed in the system. The type of information captured in this process is shown in Table 5.2.

Let me illustrate this with the imaginary 'Blossom Farm'. The site is a 10-hectare farm with sandy loam soil, on a south-facing slope. It is often dry in the summer months. The farmer in this case is a woman who loves working with children on the farm. She is highly skilled in growing fruit and likes to make preserves in the winter. She sells her produce at a local farmers' market every week. She needs to earn a living from the farm. The farm is owned by the National Trust, which has given her a secure tenancy. The National Trust wants the farm to support native bees, wants the farm to sequester carbon and will not allow any non-native perennials on the farm. On the basis of this information the functions and elements of Blossom Farm may be identified as shown in Table 5.3.

Table 5.2: Client survey tools

Process	Description	Tools / Methods
Observe the farmer	Find out about your client, even if it is yourself, and stakeholder groups.	Client interview Websites
Interview the farmer	Find out about the functions and elements they want on the farm, that is, what they want to do on the farm and what things they want.	Client interview
Resources of the farmer and stakeholder groups	Find out about time resources, timescale for project, financial resources, skill sets.	Client interview
Boundaries of clients and stakeholder groups	What are the non-physical boundaries of the site and the design? Time and money, health, cultural norms, legality, fear, skills, planning constraints.	Research with the planning authorities, and the landowners if it is a tenanted farm (on many tenanted farms the farmer is not permitted to plant trees).

2. Analysis

Once all of this information has been collected, the analysis can begin (Figure 5.4). Essentially, the interaction of the site and the farmer or grower will create a unique farm that is productive and resilient.

The site analysis starts with zone and sector mapping. A scale base map of the site will include details of soil analysis, existing trees, structures, paths of animals such as deer, water on the site such as springs and ponds, and slopes and contours. If a decision has been made to remove something, then it can be removed from the base map. A base map can be made on a computer-aided design system or on paper. On top of the base map, overlays are used to indicate the zones and the sectors of the site.

The sectors map the energy coming into the site, such as wind, sunshine, water and views. The wind mapped will be the prevailing and the cold winds. The sunshine mapping will involve the sunshine hours in the winter and summer. The sunniest sides of a site are the south-facing (in the northern hemisphere) slopes or walls; north-facing walls and slopes are colder or warm up later in the day. In the winter the latter may be a problem but in the summer or a hot climate they can provide welcome shade. Winter and summer angles of the sunshine vary, so these need to be plotted on the site to make the most of solar gain. Water enters a site in many ways: rain, streams, groundwater and mains water. It leaves via soil drainage, streams and sewerage. Retaining, enhancing or 'harvesting' a view is important to orientating a home, a bench or a working area.

The zones radiate out from the home or the barn: the areas that will be farmed or gardened intensively, those areas that will be cropped once per year, and the wilder parts of the farm.

The second process that is used is a function and element analysis (Figure 5.2; Table 5.4). A list of the functions and elements wanted on the farm is compiled, and then each element is mapped across to each function. The overarching concept is that each function is supported by many elements and each element is supported by many functions

Function	Systems of Elements	Elements
Produce food	Fruit trees Chickens Bees	Varieties of apple trees and rootstocks. Specified breed of chickens Chicken house Self-watering system for chickens Beehives for pollinators Apple press to make juice Barn for storage
Support biodiversity	Orchard sward Beehives	Sow species-rich grass beneath trees Plant bee fodder trees in hedgerow
Educational visits	Places to sit Trail Car park	Compost toilet Handwashing facilities Fire pit

Table 5.3:
Functions and elements of Blossom Farm.

Process	Description	Tools / Methods
Analysis	Teasing out what functions and elements are required on site Finding the gaps	Brainstorming and function– element analysis
	Listing all the inputs and outputs of each of the elements in the design. Do any of the outputs of one element become the inputs for other elements?	Input–output analysis
	Analysis of where the zones and the sectors lie on the site	Overlays of zones and sectors on the base map
	Analysis of how energies move about the site	Flow diagram to analyse how water, nutrients and people move around the farm

Table 5.4: Analysis tools.

to create a resilient system. At this point any gaps that arise become very clear; these represent missing elements or functions in the design.

The site analysis marks out the zones and the sectors on overlays. Zone 0 and Zone 1 are next to the entrance to the farm and further zones radiate out to the edges of the farm.

The function element analysis adds in some missing extra elements to create a resilient system.

Let us return to our imaginary farm. The fire pit at Blossom Farm can be used to burn prunings from an orchard and the resulting ash used to fertilise the orchard, supporting the function of 'producing food' as well as the function of 'educational visits'. From the sector analysis it may be inferred that water for the trees and chickens will be in short supply. Extra elements for rainwater storage and harvesting are put on the list. Keyline ploughing is added to the list to facilitate the storage of rainwater in the soil. (See Chapter 8 for more information on keyline ploughing.)

3. Design

At this point there is a series of bits of paper with functions and elements clearly written out, and a large drawing on which the zones and

Figure 5.2: Function element analysis.

Figure 5.3: Zones.

sectors are mapped out. The design process at this point is remarkably simple. Each element is placed in the correct zone and in the correct sector. It is as simple as taking the Post-it note of the element and sticking it on the base map with overlays.

Design from pattern to detail is really important when designing a large piece of land.

It is useful to start with the overview and then focus on each aspect of the design with increasing detail in each iteration of the design cycle. For instance, in the pattern iteration design of a farm a field might be marked 'orchard', and this can be integrated into the rest of the farm system, with chickens, bees, educational visits. In the second detailed iteration of the design cycle the orchard will be mapped to include the row orientation, spacing of tree varieties, choice of varieties and rootstocks and choice of sward. This enables the farmer and designer to

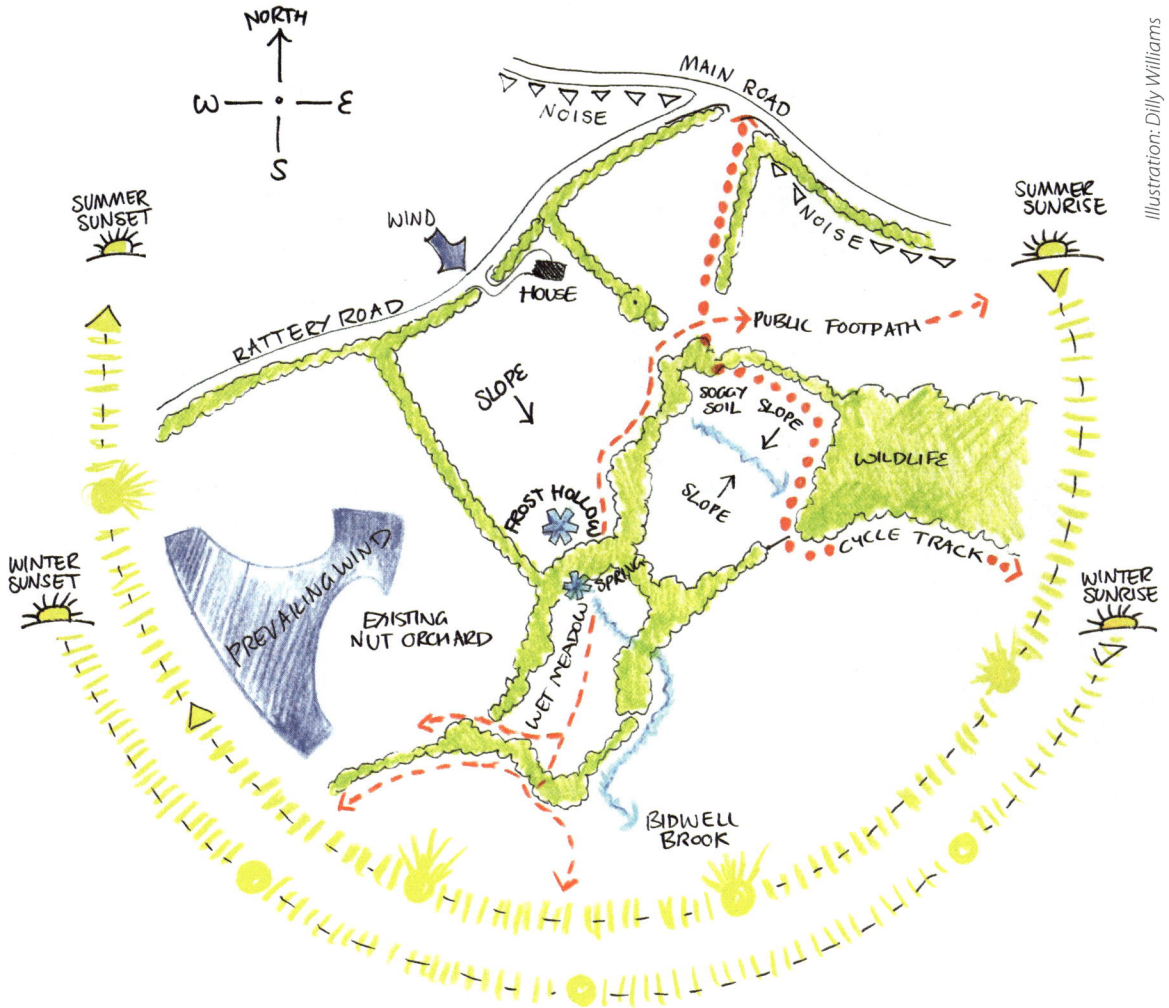

Illustration: Dilly Williams

Figure 5.4: Sectors.

create the broad brushstrokes of the design and build in the closed loop early on in the process. Designing from pattern to detail is often the hardest principle to hold on to when faced with tackling a large-scale design, but it is also one of the most important.

It is really fun to use 3D modelling to place each element in the right zone and the right sector (Figure 5.5). The use of plasticine and toys with large-scale maps drawn on the floor with

chalk, or on large sheets, allows this process to include a large group of people.

Once the elements have been put in the right zone and sector, there follows a process of 'shuffling' and fitting everything together. At this point a flow diagram can be used to ensure the flow of people and water or nutrients works on the site. To make sure the farm is as closed loop as possible an input–output analysis can be used to see how the elements link up.

Figure 5.5: Three-dimensional modelling.

What waste is produced and where? How can it become a raw material for another element? What might we need to buy in and what can we produce on site? On a farm scale this is a big piece of detailed work, especially the mapping out of the flow of water and nutrients to get this correct and working well.

The final part of the design process is to bring the list of principles and ethics to the design that is almost complete and ask if they are all fulfilled in the design. Lastly the design is thoroughly evaluated and tweaked where necessary (Table 5.5). It then evolves from paper to field.

In the example of Blossom Farm the design process places the car park, compost toilet and fire pit in Zone 0 near the barn. The fire pit is in the evening sun and wind-free sector. The log pile for the fire pit is in Zone 0 in a non-sunny area, stacked against a cold wall for extra insulation.

The orchard is in Zone 3 on the south-facing slope, and the trees are planted on the contour. The orchard is sheltered from the wind by a hazel coppice that is in Zone 4 and also provides wood for the log pile along with the fruit tree prunings. The hazel coppice supports wildlife in the form of a beetle bank by its roots; ivy is allowed to climb through the hazel trees to provide bee fodder in the winter months. The sward under the orchard is planted with deep-rooting clover and grass to feed the bees and the chickens.

The chickens have a mobile house with gutters that is moved through the orchard, but is kept as close to the barn as possible. The barn roof collects water that is gravity fed to the chickens' water trough. Some chicken feed is bought but it is organic. The chickens are kept within an electric fence that is powered by a PV panel. The bees are placed under the coppice hedge out of the wind.

The input–output analysis on Blossom Farm might look like this: The orchard produces fruit that can be eaten, sold or preserved, or used to educate groups. It produces prunings that can be burned and the resulting wood ash returned to the land as a fertiliser. Waste fruit can be left to rot for butterflies, worms or birds, or fed to the flock of chickens, or composted to create more fertility. The trees themselves soak up carbon and provide shade for chickens. The chickens provide more fertility for the trees, eat the pests and lay eggs.

Beehives provide pollination; honey is extracted for sale. The wild flower meadow underneath the orchard provides extra bee fodder and a place for wildlife and educational opportunities for school visits. On Blossom Farm the flow of the water through the site is slowed down, and water is stored in rainwater collection tanks. It is also stored in ground that has been keyline ploughed so that water will penetrate deep into the soil. The trees have been planted across a slope, or along the contour, and all of this has cut down on the floods that happen every winter at the bottom of a hill. The water is very important on the farm for wildlife purposes and water management is key to productivity (see Chapter 8).

4. Implementation

Once the design is complete the implementation plan can be created; this often spans two to three years. Permaculture-designed farms are complex to set up and require a lot of work and energy at the beginning. However, once everything is set

Process	Description	Tools / Methods
Design	Placement of the elements into the design	Base maps and overlays Random assembly Wild design Planning for real
	Integrating elements to create a fully functioning design	Input–output analysis Flow diagram
	Making sure it works	Check alignment with principles and ethics and the farmer's feedback Produce a client report, drawings and implementation plan Business plan and costings
Evaluation and tweak of design		Action learning cycle

Table 5.5: Design tools

How a permaculture-designed orchard differs from an industrial orchard and an organic orchard

In an industrial orchard a monocrop of dwarf apple trees is planted at 25,000 trees per hectare, the grass is sprayed off beneath the trees and the trees are fertilised with bought-in fertiliser. They are sprayed 10–14 times per year to control scab and any pest attacks in order to create a blemish-free apple. They have to be irrigated occasionally owing to climate change. The apples are picked and put in cold storage with gases added to prolong their life. They are then wrapped in plastic and sold into a supermarket, either in the country of origin or elsewhere around the world.

In an organic orchard the fruit trees may be two or three varieties of one type of fruit, planted at 500 trees per hectare. The grass is cut with machinery under the trees, which may be sprayed with sulphur and liquid seaweed to create a better skin finish on the fruit. The fruit is sold into a large box scheme or to supermarkets, and blemished fruit is made into juice.

In a permaculture orchard like Blossom Farm the trees are chosen to be genetically resistant to disease; there are 10–20 varieties of mixed types of fruit and these are planted at a density of 500 per hectare. They are planted on the contour of the slope with wide row spacing. The trees do not need irrigation, since keyline ploughing was carried out when they were planted; this allows rainwater to percolate deep into the soil. Each tree is mulched and underplanted with plants that attract predatory insects that eat the pests. A flock of chickens graze the grass clover and herb-rich sward; these feed the trees and break the pest cycles and provide extra income for the farmer. Gutters on the barn and chicken houses collect rainwater, which automatically fill troughs for the chickens. Polycultures of fruit are planted, providing sequential cropping and increasing diversity, so lessening the movement of disease through the trees. Fruit is picked little and often and sold locally. Blemished fruit is made into juice or fed to the chickens or left on the ground to support biodiversity. Butterflies in particular enjoy rotting fruit. Educational groups of children come to the orchard regularly and pick some apples and make juice. They often light a small fire from the prunings and the ash is fed back to the trees. This provides extra income for the farmer, who also really enjoys these visits.

Although industrially grown apples have the greatest overall yield, this system also requires the greatest amount of capital and energy input in the form of fossil fuel, human input, work and money. It also externalises the costs of pollution and climate change, and the loss of biodiversity, and human illness. Meanwhile the permaculture farmer works four days per week and employs part-time people to help her out. She does not play the guitar but uses her spare time to be with her family and friends.

up, the maintenance of the site becomes much easier and self-regulating. The detailed planning in the design process makes the creation of an implementation plan and business plan very straightforward.

Once the design starts to move from the

page to the field, it usually requires 'sculpting' to fit the physical site and at this point another round of 'evaluation and tweaking' goes on as the design becomes a reality. This process is called 'action learning'. It is learning by doing. As many permaculture-designed farms are pioneering, there is rarely a textbook to follow.

The skill required to implement a design, especially a large-scale design, should not be underestimated. A complex set of skills are required that are rarely found in one person and may require a team or bought-in expertise for some parts of the implementation process.

In the Blossom Farm example it was quite straightforward to write on a piece of paper 'chicken tractor' or 'sequential harvested fruit orchard with wild flowers', but at the implementation stage the farmer requires skills, or needs to buy in skills, to design and make a mobile chicken house and to work out how to harvest rainwater for the chickens, what green manures to choose, what fencing system to use, what breed of chicken to choose. The orchard requires the choice of sequential fruit on the correct rootstock for the soil type and local climate, and the correct fertility, windbreaks and pollinators. The success of the system lies in these decisions being more or less correct. Finally, if the farm is to produce income then the produce needs to be sold and business plans will be needed.

5. Maintenance

Mollison suggests that a good design will be self-regulating and will evolve. On a larger scale, permaculture farms are very productive and require lower labour inputs.

Do permaculture food production systems work?

There is very little research on permaculture farming systems. On the basis of the permaculture design methodology it may be expected that these farms will have soils that sequester high amounts of carbon and are resilient to climate change because of the high level of mulch used and the polycultures grown, and that they will be highly productive per hour worked.

The few pieces of scientific research that have been done specifically on permaculture systems demonstrate just that. Rebecca Laughton's research paper 'A matter of scale' in 2017 showed that small-scale permaculture, agroecology farms and market gardens were more productive than the industrial system of farming. They had higher yields per unit area and per hour of work. They demonstrated a significant reduction in inputs and waste products, and income and the number of people employed were higher per unit area. These farms also motivate the provision of social and environmental benefits, extra care being taken with the use of water resources and to support biodiversity.[15]

Ferme du Bec Hellouin is a permaculture-designed farm in France which has been carrying out scientific research in partnership with universities. It has been demonstrated that the farm has exceptionally high yield and productivity both per unit area of land and per hour of work. It has also created soils very high in organic matter. The farm has demonstrated that it is resilient to climate change because of the high diversity of crops and the effect of high levels of SOM and mulching. The farmers reported that the farm was full of biodiversity.[16]

What do permaculture food production systems look like?

Although the number of permaculture farms is growing fast, they add up to a tiny percentage of global agriculture. Many permaculture systems have the core aim of self-reliance, that is, that people can 'make a living from products derived from stable landscapes, although this is not the primary aim of permaculture, which seeks first to stabilise and care for land, then to serve household, regional and local needs, and only thereafter to produce a surplus for sale or exchange'.[17] Thus it is common to find permaculture projects in intentional communities that want to create food-rich productive areas to feed themselves, or in the form of small-scale community gardens in urban areas.

The most well-known permaculture farms are Sepp Holzer's farm Krameterhof in Austria,[18] Organic Lea in London,[19] and Ben Falks' Whole System Research Farm in Vermont.[20] Geoff Lawton's Zaytuna Farm in Australia is home to the Permaculture Research Institute; he has also developed a four-hectare farm in Jordan just a few miles from the Dead Sea, through his 'Greening the Desert' project.[21] Most of these farms have taken marginal, degraded land and completely transformed it into a productive food-bearing and biodiverse landscape.

We can see permaculture at work on a large scale in a multitude of small projects in Cuba. In the early 1990s Cuba went through what Cubans refer to as the 'Special Period' when their oil supplies were reduced drastically because of the collapse of the USSR. Essentially, Cuba went

through a rapid 'peak oil' transition. Food was rationed and the ability to drive cars to buy food was dramatically reduced. The wonderful documentary *The Power of Community* details how Cuba created small-scale, relocalised, low-carbon food systems using permaculture design methods.

Today an estimated 50 percent of Havana's vegetables come from inside the city, while in other Cuban towns and cities urban gardens produce from 80 percent to more than 100 percent of what they need. The result: more self-sufficient communities which consumed more fruits and vegetables were created; petrochemicals were replaced with bio-fertilisers and bio-pesticides and oxen were enlisted as an alternative to machinery reliant on petroleum.[22]

This quote is from 2006; since then oil supplies into Cuba have resumed and these figures have changed. However, the Special Period demonstrated that, when needs must, new relocalised food systems can be implemented in just a few years.

Huxhams Cross Farm case study in Chapter 10 illustrates in more depth how a farm can be designed using the permaculture methodology.

Case study:
Organic Lea, London

Organic Lea is a workers' cooperative based in northeast London at the Hawkwood Nursery in the Lea Valley, which was the main food-producing area for London until the 1970s (Figures 5.6 and 5.7).

It started with a group of permaculturists creating a market stall at the Hornbeam Centre on a busy street in Walthamstow, East London, by sourcing food from small organic growers in and around East Anglia and surplus food from local allotments and people's back gardens. In 2009 they took on a ten-year lease on a five-hectare disused plant nursery in the Lea Valley. A permaculture design was carried out and now the place is a thriving centre of food production and training.

The core of the farm comprises a large glasshouse with a water-harvesting system, where all year round diverse salad crops are grown, together with two hectares of raised beds. These are surrounded by an orchard and vineyards. The farm is nestled on the edge of Epping Forest; the whole site, apart from the entrance, is surrounded by woodland. It is approached through suburban Chingford. The site is registered as organic and is also vegan, with no animal inputs.

The food is sold locally through the Hornbeam Cafe market stall and via their vegetable and fruit bags and delivery scheme. It also goes to cafes and restaurants. The cooperative top up their own produce, buy in produce from organic farms within a 95 km radius and offer a Crop Share scheme whereby excess vegetables and fruit from local allotments and gardens can be sold via the market stall.

Organic Lea positively supports biodiversity and in particular works to support wild bees and other pollinator insects by providing habitat and food on its site and providing training in North London for others to do so. They have top bar beehives on site and occasionally harvest some honey. These hives allow the bees to keep their own honey in the winter instead of being fed with sugar.

Figures 5.6 and 5.7: Organic Lea glasshouse.

Organic Lea runs training courses and provides volunteer opportunities to help to create a vibrant food community in North London. It holds regular open days and is part of a thriving movement in London to produce local food in the city.[23]

Case study:
La Ferme du Bec Hellouin

Started in 2006 by two novice but enthusiastic gardeners in Normandy – Charles Hervé-Gruyer and Perrine Hervé-Gruyer – this 1.8-hectare garden was created using permaculture principles and practices (Figure 5.8). In addition they used traditional Parisian 19th-century market-gardening methods, called 'biointensive gardening', interwoven with Eliot Coleman's 21st-century flatbed horticulture system from the US. They call their approach 'micro gardening'. They grow fruit, vegetables and aromatic and medicinal plants. They have two garden islands in a stream that runs through the site, which allow them to grow an even wider variety of crops because of the islands' microclimates. In the ponds they produce carp. Reeds from the ponds go to make compost that enriches the soil. They have 600m² of glass, a small forest garden and a mandala garden all contained within the 1.8-hectare site. The farm is in the Bec valley, not a traditional area for market gardening with its thin soils over flint and cold, late frosts. On a neighbouring three-hectare site the Hervé-Gruyers have a small pasture with donkeys, draft horses and sheep and another edible forest garden. They also have 12 hectares of woodland.

Research on this site has shown that the soil in the beds is higher in SOM than are industrially farmed soils, with improved soil aggregates and greater nutrient availability. This should not be a surprise, given the amount of compost and manure added to the intensive beds. The yields are exceptionally high for such poor soils and such a small space, and the biodiversity has gone through the roof.

The Hervé-Gruyers can earn an average of €900–1,500 per month from 1,000m² of growing

Figure 5.8: La Ferme du Bec Hellouin.

space on an average of 43 hours work per week. This reflects exceptionally high productivity in terms of both yield and work hours.[24]

The Hervé-Gruyers offer tours and training. Their tiny farm has caught the imagination of the French and many films have been made about it. They have written a book and made a film called *Miraculous Abundance* and are now focusing on scientific research on the farm to better understand what it has to offer.[25]

Case study:
Regional permaculture projects in the global south

The following case study of the Himalayan Permaculture Project was the result of a special research project conducted by Dr Anne-Marie Mayer in Nepal in 2018. Mayer is a nutritionist who works for the FAO across the global south and is also trained in permaculture.

The Himalayan Permaculture Centre (HPC) is a grassroots NGO set up in 2010 by trained farmers from Surkhet district in Mid-Western Nepal, to implement sustainable rural development programmes in conjunction with Chris Evans, a UK-based permaculturist. It operates in two rural districts of Nepal – Humla and Surkhet – where the project maintains resource centres for demonstration and teaching with 850 directly participating households. The annual budget of the HPC in the two districts was approximately £84 per household in 2018. The HPC builds on experience gained over the past 30 years in training in and implementation of permaculture design and practice in remote and challenging environments in Nepal. The

HPC aims to build sustainable agriculture and resilient domestic food and energy security for the communities living in the project area. Work is organised in five areas: food security, health, education, livelihoods and capacity-building.

The HPC has a strong potential to improve nutrition. Multi-sector projects such as the HPC require only a small budget and have the potential to deliver wide benefits. The longevity of the HPC relates to low costs and the capacity that has been built among local people to implement the project.

By means of permaculture design, integrated systems that cross traditionally separate sectors can be used to build the health and nutrition of communities in a socially, economically and environmentally sustainable way. With the multifaceted challenges that face rural Nepalese communities, a truly multi-sectoral programme approach is essential and the HPC offers a model. HPC staff and Barefoot Permaculture Consultants are already working as ambassadors for permaculture and the work carried out by these consultants in response to the 2015 earthquake is evidence of the value of the HPCs work.[26]

Over 45 techniques are recommended by which farmers may strengthen household resilience by increasing and diversifying farm productivity; reducing the cost of domestic activities in terms of time, labour and money; improving health through better nutrition and hygiene; using and recycling local resources and protecting the environment. The techniques promoted are described in the HPC farmers' manual. The HPC operates through (i) demonstration of techniques, (ii) training of communities, (iii) provision of resources and (iv) research on techniques and approaches. Activities are supported in the communities through capacity-building by HPC technicians and Barefoot Permaculture Consultants – farmers who have been selected by communities for their leadership, technical knowledge and successful application of techniques.

The HPC is a truly multi-sector programme with activities and planned outcomes that cross the disciplines of agriculture, livelihoods, health and environment, among others. The strength of the HPC is in the connections between the different activities and the ways in which they are integrated. Multiple elements are in place for key functions, and each element serves multiple functions. In this way communities build resilience. This is how permaculture designs work to address the multiple challenges and needs of communities in a participatory way.

From a nutrition perspective, therefore, many elements are in place to support adequate food and nutrition security and the HPC is a good case study to demonstrate the potential for permaculture to support this.

There are several clear links between HPC strategies and activities and potential positive impacts on human nutrition:

1. Building resilient and productive agriculture systems

There are several challenges to farming in the area, including drought, pests, lack of market access, and migration. To meet these challenges the HPC has introduced diverse, perennial agriculture systems with a rich mix of species in agroforestry systems using a range of propagation techniques. Improved livestock breeds have been introduced, and stocks have been reduced to allow natural forest regeneration on hillsides and also to reduce labour demands. Composting and water systems are managed to support increased production.

Novel systems and techniques sit alongside the traditional farming systems. A 'System of Rice Intensification' has been introduced which increases yield using local fertility resources.

Participating farmers report improved productivity, but yield data are not available to verify this.

2. Improving livelihoods

The introduction of agricultural diversity is a strategy to improve diets, resilience and livelihoods. The HPC also supports cottage industries and herb-processing enterprises. Income derived from livelihood activities and money saved from adoption of HPC techniques have been used for purchases of relevance to nutrition in the short or long term. For example, the payment of school fees has potential long-term benefits for nutrition. Purchases of soap, water filters and medicines also have potential benefits for nutrition through improved health. All of this demonstrates how the increased productivity accomplished by this project increases income.

3. Diets

Most participants reported increased consumption of fruit, vegetables and pulses since joining the HPC. For some households there was a reduction in the consumption of animal milk owing to a shortage of labour for livestock rearing, but the mitigation measures put in place (such as fodder supply, improved breeds, improved water supply) have minimised this effect. Other challenges for nutrition include loss of nutrients from overcooked vegetables and overprocessed rice. Neither of these habits is easy to change. With the additional income available to households, there is also a risk of increased consumption of overprocessed foods where these are available from local shops. Efforts have been made by the HPC to value traditional food culture and resist the introduction of overprocessed foods. Cooking demonstrations, food processing, food combining and the use of locally developed complementary food (for example, sarbottam pitho) help to reinforce local traditional practices, but widespread use of bought processed foods was reported by respondents and HPC staff.

4. Gender empowerment

The HPC is approaching gender empowerment from different angles. There are many ways in which time and drudgery are saved through HPC activities. These include the use of special technologies such as energy-efficient stoves, but what is clear is the link between improving the sustainability of natural systems and the time needed to provide for human needs.

Water systems, soil fertility improvements, reforestation all connect to time-saving for women. This has not been widely recognised in the debates on women's time and nutrition and is a key finding. For nutrition, there are clear links between women's workload and nutrition. From the testimonies of women interviewed:

- Women are able to spend more time with children.
- Women's own energy expenditure through long hours of drudgery has decreased.
- House cleanliness and hygiene are improved.
- There is more time to prepare special meals for children.
- There is more time for income-generating activities or social enterprises.
- There is more time for attending meetings and socialising.

This illustrates how a well-designed permaculture system is very productive, and that 'everything gardens'; in this case there is no playing of guitars on the veranda, but more meaningful time is spent with the family and generally having fun.

5. Nutrition and health promotion

Nutrition and health promotion are among the functions of HPC, including nutrition training, cooking demonstrations and health fairs. The farmers' manual includes a chapter on 'Diet and nutrition', which describes the nutritional properties of foods, food processing, production of sarbottam pitho and care of the sick or malnourished child. This demonstrates the ethic of people care in this project.

6. Maternal and child care

Adequate maternal and child care is critical to the prevention of malnutrition. The global recommendations are for women to increase their consumption of nutrient-dense foods and to take more rest during pregnancy. The HPC contributes to the improvement of maternal care by making nutritious foods available and by reducing women's workload. This demonstrates the ethic of people care and fair share.

7. Infant and young child feeding (IYCF) practices

HPC activities that save time and encourage rest-taking during the critical 1,000 days (pregnancy and the first two years), together with the greater availability of diverse foods, are important for IYCF. Increasing the diversity of crops is a permaculture principle.

8. Improved water and sanitation

The HPC has introduced many hygiene promotion activities and easy-to-implement technologies to ensure food safety and hygiene practices. The provision of piped water to households and sanitation are key benefits for health. These improvements in food hygiene demonstrate the application of the people care ethic to this project.[27]

Training in permaculture food production systems

Permaculture is a well-established international movement. There is a standardised 12-day course called the Permaculture Design Certificate (PDC), which can be found across the globe. Permaculture design courses are run as a full-time 12-day course, part time, online, or as evening classes; they are all a minimum of 72 hours. They are sometimes delivered on permaculture farms or projects and sometimes not. Once the PDC has been completed, there is a path on to a diploma. The Permaculture Diploma takes a minimum of two years, depending upon the apprentice, and requires them to complete 10 designs. Once those are complete, the apprentice becomes a designer and is permitted to run the PDC. There are many permaculture designers. Some specialise in garden-scale designs, some on people and organisation design, and some on large-scale farm projects.[28] Permaculture is rarely offered in state education.

Where is the permaculture food production movement going?

The permaculture movement has regular global, international, national and regional gatherings. It has very good networking systems and a great number of websites that list the permaculture projects across the globe. There is a huge range of films and podcasts and written information about permaculture and how to implement it into your life or food system.

The public face of the permaculture movement is provided by the Permaculture Institute in New Mexico,[29] the Permaculture Research Institute in Australia[30] and the Permaculture Association in the UK.[31] *Permaculture* magazine is published quarterly in the UK and in the US as well as online.[32]

In the UK, the Permaculture Association has three main charitable aims, namely, to advance public education, to undertake and share research and to conserve the environment. There are lively educator groups and initiatives, research meetings and projects and a demonstration network called the LAND network.[33] Members work together in local, regional and national groups and the movement as a whole embraces and values diversity and

The Story of Hodmedods, Grown in Totnes, and the decommodification of grain

After World War II the perceived need to increase the production of basic foodstuff by using nitrogen-based fertilisers and pesticides was acted upon across Europe and the US. The industrialisation and commodification of grains and pulses saw new varieties bred that were dependent upon pesticides and high nitrate inputs. These were called higher-yielding varieties (HYVs) and were the precursors of genetically modified organisms (GMOs). Milling and other processes such as the de-hulling and decortication of beans were scaled up to take place on a large scale in factories. Small-scale mills and processing plants went out of business. To stay in business, farms needed to get bigger and have higher yields. In these large-scale industrial systems, once a grain or pulse crop has left a farm it is difficult to track its destination to its final market, since it has become a commodity within a long supply chain.

The grain may end up in a cattle feedlot in China or in a supermarket sliced bread loaf.

The Transition food groups that started up in 2010–15 carried out food audits of local food supplies. They realised that in order to reduce the carbon footprint of our food we need to eat more local and plant-based diets. In the UK most of the pulses grown, mainly fava beans, are exported; the UK imports chickpeas, kidney beans and soya beans. In the early days of the Transition movement many local food groups mapped their local food supplies and found that it was difficult to buy local pulses, oats, wheat or other grains.

Hodmedods grew out of Transition Norwich, and out of permacultural thinking. They pioneered the relocalisation of grains and beans in the UK. They initially asked the huge factories in East Anglia to sell them small amounts of pulses. These were packaged and sold online via a website or wholesale into local shops. They are now commissioning farmers to grow small-scale grains, peas and beans and are recreating the processing practices required to process them on a small local scale. They are also reintroducing crops such as lentils into UK farms.[35]

In Totnes, Holly Tiffin and her Grown in Totnes team mapped out the food being produced within a 50km radius of Totnes. Although lots of local vegetables, milk and meat could be found in the shops, you couldn't buy local flours or oats – all of which were grown abundantly in the region. Grown in Totnes had an even more ambitious aspiration: to relocalise cereal production, namely, oats and wheat, to within a 50km radius of Totnes. In order to get oats or flour from local farms into the kitchens of Totnes the infrastructure and skills of small-scale mills had to be recreated and the skills of milling relearned. This made relocalising production not only difficult but economically challenging.[36] Grown in Totnes has now developed into Dartington Mill

Continued following page…

Continued from previous page…

(Figure 5.8), which buys and processes small-scale local grains, for making into a 'Local Loaf' and for sale as flour. Dartington Mill is owned by two farm enterprises and a bakery. One of these farms is the Apricot Centre at Huxhams Cross Farm.[37]

The varieties of wheat being grown for both Hodmedods and Dartington Mill have been specially bred for these organic local small-scale systems. One of these varieties, YQ (Yield and Quality) wheat, was bred by Professor Martin Wolfe in collaboration with the John Innes Institute and the Organic Research Centre in the UK. He took 20 modern wheat varieties and crossed them; the resulting seeds were saved and resown, bulked up and saved again, resulting in a 'population wheat'. This term refers to the huge genetic diversity contained in the genotype of this type of wheat, which makes it very adaptable to different soils and climates and resilient to pests and diseases.[38]

John Letts, a farmer and plant breeder in the UK, has spent years recreating the pre-1940s landraces of wheat by visiting ancient thatched cottages and finding ears of wheat left in the thatch which were still viable, that is, would still germinate. These types of wheat are tall and deep rooting and have strong flavours. They have been bulked up and are now grown as 'heritage wheat' – a form of landrace population wheat – and popular with permaculture farmers.[39] Many farmers all over Europe and the US are now pioneering and developing new population wheats, since they are deep rooting and remarkably weed free and resistant to disease and have reliable cropping, whose yields do not fluctuate much from year to year, even in erratic weather conditions caused by climate change.

Josiah Meldrum

Figure 5.9: *Hodmedod's new American stone mill with miller, Keith Malcolm.*

Kimberly Bell of the Small Bakery in Nottingham has pioneered baking and bread-making with flour made from the YQ population wheat. The flour is naturally low in gluten, so sourdough methods are used to give the rise and they make delicious breads and pastries. At annual meetings of the Grain Lab, bakers, farmers, scientists, millers and nutritionists discuss and recreate the new decommodified and relocalised grain systems.[40]

In the US this grain culture is thriving with the creation of new mills. The New American Stone Mill in Vermont is building mills that do not overheat the grains, thus preserving the full nutrients and taste in the flour for the consumer.[41] Maine Grains set up a milling business in Maine which has changed the local economy by bringing together farmers, bakers, brewers and consumers in a short healthy food chain. They offer a model of how to restore local grain heritage.[42] You can browse the website of the North East Grain Shed to see what is happening in the US and that of UK Grain Network to see what is happening in the UK.[43]

as such is an incredibly diverse community of people who attend and host gatherings, training and events throughout the year. Although there are not many permaculture farms as such, the permaculture movement has been very influential through its principles and thinking about food systems. This is illustrated by the story of the relocalisation and decommodification of grains in the UK and the US. The permaculture movement was also key in the formation of and is part of the toolkit of the regenerative agriculture movement.

The Transition Towns movement arose out of the permaculture movement. Rob Hopkins was running permaculture design courses in Ireland and started to design 'energy descent' plans for whole communities with his students. He then moved to Totnes and met Naresh Giangrande and together they created a dynamic movement that now spans the globe. These communities are actively engaged in creating pathways towards low-carbon resilient systems for food, energy, transport and more.[34]

The Transition Towns movement has created 'food groups' around the world that are examining the food systems that are currently in place and actively working on creating sustainable food supplies into cities and towns. This has in turn inspired the creation of pioneering not-for-profit companies such as Hodmedods. ∎

In a nutshell

- Permaculture is a design system used to create resilient and sustainable systems.

- The ethics of earth care, people care and fair share are at the core of these systems.

- SADIMET (Survey, Analysis, Design, Implementation, Maintain, Evaluation, Tweak) is the main methodology for creating systems from tiny gardens to large-scale farms.

- Permaculture-designed systems aim to be closed loop and to produce food, and often contain more perennial crops than most farms.

- They are very productive and offer many yields besides that of the crops, with improved quality of life for those involved in them.

- They support biodiversity and often restore local water cycles and sequester large amounts of carbon.

- Permaculture is a key component of the regenerative agriculture movement.

Further reading, links and films

Aranya, *Permaculture Design: A Step by Step Guide*, Permanent, East Meon, UK, 2012.

L. Macnamara, *People and Permaculture*, Permanent, East Meon, UK, 2012.

R. Morrow, *Earth User's Guide to Permaculture*, Permanent, East Meon, UK, 2010.

R. Perkins, *Regenerative Agriculture*, self-published, 2015.

M. Sheppard and A. Lappe, *Restoration Agriculture: Real World Permaculture for Farmers*, Acres US, Greeley, CO, 2014.

P. Whitefield, *The Earth Care Manual*, Permanent, East Meon, UK, 2004.

Becky Hosking's BBC documentary *Farms of the Future*. The story of the film and the link to it can be found via her website: https://www.rebeccahosking.co.uk/a-farm-for-the-future.

Permaculture magazine has made a series of *Living with the Land* short films: https://www.permaculture.co.uk/living-with-the-land

There are many permaculture films, from classics such as *In Grave Danger of Falling Food* with Bill Mollison; Inhabit, visits to Geoff Lawton's 'Greening the Desert' project in Jordan and Sepp Holzer's farm in Krameterhof in Austria; and *Green Gold*, John Liu's story of reclaiming the Chinese Loess Plateau. Available via: https://filmsfortheearth.org/en/issues/permaculture

For listings of films about regenerative agriculture and permaculture in New Zealand, Australia and the US, see: https://happenfilms.com/permaculture

1 B. Mollison, *Permaculture: A Designers' Manual*, Tagari, Tyalgum, 1988, p. 2.

2 Ibid., p. 36.

3 Ibid., p. ix.

4 Ibid.

5 Ibid.

6 J. Macy, *The Work That Reconnects*, New Society, Gabriola Island, BC, 2014.

7 D. Holmgren, *Permaculture: Principles and Pathways Beyond Sustainability*, Permanent, East Meon, UK, 2002.

8 Mollison, *Permaculture*.

9 https://www.footprintnetwork.org

10 Mollison, *Permaculture*.

11 P. Wohlleben, *The Hidden Life of Trees*, HarperCollins, London, 2015.

12 I. Tree, *Wilding*, Picador, London, 2019.

13 Mollison, *Permaculture*, p. ix.

14 Ibid.

15 R. Laughton, 'A Matter of Scale: A Study of Productivity, Financial Viability and Multifunctional Benefits of Small Farms 20 Hectares and Less', *Landworkers' Alliance*, 2017,

16. K. Morel, C. Guegan and F. Leger, 'Can an organic market garden without motorisation be viable through holistic thinking? The case of a permaculture farm', Ferme du Bec Hellouin, 2015.

17 Mollison, *Permaculture*, p. 18.

18 S. Holzer, *Permaculture*, Permanent, East Meon, UK, 2004.

19 https://www.organiclea.org.uk

20 B. Falks, *The Resilient Farm and Homestead*, Chelsea Green, Hartford, VT, 2013.

21 https://www.geofflawtononline.com

22 *The Power of Community: How Cuba Survived Peak Oil*, film: https://www.filmsforaction.org/watch/the-power-of-community-how-cuba-survived-peak-oil-2006

23 For more information: https://www.organiclea.org.uk

24 Morel et al., 'Can an Organic Market Garden …?'

25 C. Hervé-Gruyer and P. Hervé-Gruyer, *Miraculous Abundance*, Chelsea Green, Hartford, VT, 2016.

26 http://himalayanpermaculture.com

27 Ibid.

28 https://www.permacultureassociation.org.uk

29 https://www.permaculture.org

30 https://www.permaculturenews.org

31 https://www.permacultureassociation.org.uk

32 https://www.permaculture.co.uk

33 https://www.permaculture.org.uk

34 https://www.transtionnetwork.org

35 https://www.hodmedods.co.uk

36 https://www.transitiontowntotnes.org

37 https://www.reclaimthegrain.uk

38 https://www.wakelyns.co.uk

39 https://www.bakerybits.co.uk

40 https://www.Smallfoodbakery.com

41 https://northeastgrainshed.com

42 https://www.newamericanstonemills.com

43 https://northeastgrainshed.com/; https://www.ukgrainnetwork.com

Chapter 6

Agroforestry food production systems

… by the time the oaks were 10 years old and taller than both him and me. They were an impressive sight… his forest measured eleven kilometres across at the widest point… when you remembered that it had all emerged from the hands and spirit of this one man without any technical aids, you saw that men could be as efficient as God in other things beside destruction.[1]

Jean Giono

Introduction

Trees have been systematically lost across the world as forests have been cleared to obtain their timber and to make way for industrial farming. Although 30 per cent of the world is still forested, forest is being lost at an alarming rate; 17 per cent of the Amazon rainforest has already been cleared. Professor Martin Wolfe suggests that 10,000 years ago there were approximately 1.5 million trees per person in the world whereas today there are approximately 400 trees per person.[2]

European countries lost most of their original forests hundreds of years ago. The loss of trees has increased as most of the farmed landscape has been industrialised over the last 100 years. In the UK, where more than 70 per cent of the land is farmed, the RSPB estimates that 50 per cent of the hedgerows have been taken out to make way for larger and larger fields and machinery since World War II. Only 13 per cent of the land in the UK is forested. Across Europe the figure is 30 per cent. The same process is happening across the world.

Forests are being cleared in Indonesia in favour of palm oil plantations. Palm oil is used in many processed foods. The Amazon rainforest is being cleared in favour of soya bean production for animal feed and oil production. Canadian and Scandanavian old growth native forests rich with biodiversity have been cleared and replaced with monocultures of pine forests.

The reintroduction of trees into the farmed landscape is called 'agroforestry'. This chapter will describe the principles and practices of agroforestry and the benefits and challenges of putting trees back into the farmed landscape. It includes three case studies of agroforestry in practice and information about where to find out more about this system.

What is agroforestry?

The term 'agroforestry' was coined in 1977 to describe the combination of agriculture with forestry. Agroforestry has in reality been

Photo: John Morrison, Alamy

*Figure 6.1:
Dedham Vale: ancient
pollarded willow on
pasture that is also
a flood plain in the
Stour Valley.*

practised for millennia. It is not necessarily organic, but can generally be considered a sustainable practice. Agroforestry has no charismatic founder as with biodynamic farming or permaculture; it has arisen out of agricultural research and academia, and indeed out of age-old pre-industrial farming practice.

As the name implies, 'agroforestry' is a combination of a forestry crop, which provides a yield and/or an environmental service, with an agricultural crop. The combination increases the overall yield from the land. The aspect of growing trees is integrated throughout an agroforestry system and can provide many benefits as well as challenges. Agroforestry changes the way the farming is done, and the size of machinery that is used, and can be customised to many different contexts, farming systems and climates.

Agroforestry systems are found in traditional farming practices across the world and throughout history. The landscapes containing ancient agroforestry systems are often cherished ones and therefore protected. Modern agroforestry designs are often implemented as an aspect of permaculture design and of biodynamic, organic or regenerative farming.

Where did agroforestry start?

Many ancient culturally rich agroforestry systems are found across the world. The *dehesas* of Spain are beautiful permanent pastures dotted with evergreen oaks used for cork production, and grazed with cattle. In Greece and Italy small-scale olive groves undercropped with wheat are common.

It is estimated that 14 per cent of the farmed

land in Europe is already farmed in a form of agroforestry. The majority of this is found in the south of Europe, where the benefits of adding trees to the landscape are the highest, providing shade and pulling water up from deep underground. The light intensity in the south is higher, so there is less concern about shading out the crops and pastures grown beneath trees. In Northern Europe, with its lower light levels, traditional agroforestry systems are less common, but we still find cider and perry orchards in the UK undergrazed by sheep or cattle.

Parkland pastures are commonly found surrounding grand houses, grazed with sheep or cattle, who receive shade from the trees. In Dedham Vale and the Somerset Levels pollarded willow is grown at the edges of rivers to provide willow for charcoal or basket making, while the meadows are kept free for grazing (Figure 6.1).

Oliver Rackham describes ancient systems of woodland pastures in his classic book on woodlands; in these systems the word 'pannage' describes the rights of local people to graze livestock on the acorns or beechmast.[3] They are some of the finest landscapes in the UK and are often protected and preserved in Areas of Outstanding Natural Beauty (AONBs).

Such systems were developed because of their multifunctionality. For example, a cider crop is harvested over a period of a few weeks each year, and willow only needs to be harvested once per year. The cider apple crop is grown on standard trees; that is, the crop is two to three metres from the ground, which allows the grass in the same space to be used for grazing sheep.

More complex forms of agroforestry or home gardens are traditionally found in hotter climates with higher light levels. These occur in Kerala (in India), Nepal and Indonesia. In Mexico they are called 'family orchards' and are thought to have persisted since Mayan times. Research now shows that Amazon forest, before colonial times, was 'farmed' in the sense that edible trees and medicinal plants were planted in areas near villages and along tracks and paths through the forest. The soil was stabilised and improved with biochar – charcoal that can be added to compost heaps and loaded with nutrients and then added to the soil. In parts of Amazonia 'terra preta' or 'black soil' are found in areas that were inhabited before colonial times; these are thought to show where the equivalent of biochar was used to stabilise and add fertility to soil.

Such traditional agricultural systems supported large populations of people, but have been eroded over the last few centuries. Some of the old systems have been preserved, but it has been rare until recently for new ones to be established. Agroforestry has been updated for the purposes of modern food production systems and the benefits of it are so great that it is undergoing a revival across the world.

To give an example, currently in the global south, coppice or pollarded tree systems are frequently used in agroforestry systems. The FAO describes an agroforestry system for Sub-Saharan Africa which incorporates leguminous trees that fix nitrogen and provide fodder for cattle, whose manure will return to the soil. In between the trees, crops of pulses are grown to provide alternative sources of protein. In some places the trees are harvested to provide fuel for cooking. This makes the harvesting of firewood easier and safer for the women who do this work, and the fuel comes from a sustainable supply rather than from cutting down forests. The women often plant medicinal plants in these complex agroforestry systems, such as neem in India or moringa in Africa.[4]

Principles of agroforestry

Agroforestry employs two main principles: increasing diversity in the food system and the multifunctionality of the farming systems. When combined, these two principles increase the resilience of the food production system.

The diversity and resilience of the food system are maximised by increasing the number of crops in both the tree layer and the arable crop layer. The tree layer itself may comprise many different crops, or different varieties of the same crop. For example, different types of fruit trees may be combined, or different varieties of apples may be grown. At the arable cropping level the same principle applies: diversity can be increased by increasing the number of crops, or the number of varieties of the crop plants. At a genetic level the diversity of the crops and trees can be increased by choosing trees that are grown from seed and cross-pollinated, by using landraces or population grain crops or open-pollinated vegetable crops. In livestock systems resilient breeds and hybrids of livestock can be chosen. Increasing the diversity of the food systems makes them more resilient, stable and productive.[5]

Multifunctionality also increases resilience in agroforestry systems. A tree crop can provide a crop as well as a service. The introduction of a nut tree, for instance, will provide nuts, shade and a windbreak for animals grazing beneath it.

Practices of agroforestry

Modern agroforestry falls into six main categories: alley cropping, forest gardens, woodland pasture, parkland, shelter belts and riparian buffers.

1. Alley cropping

Alley cropping, as its name suggests, combines vertical alleyways of tree crops with a horizontal area that is used for either grazing or a crop. This increases the yield from the total system, but the tree crops do not take up much space from the other areas. There are two main types of alley cropping: silvopastoral systems, where the trees are combined with animals, and silvoarable systems where the trees are combined with arable cropping areas.

Figure 6.2: Eastbrook Farm dairy cows in alley system.

Farm Photo: Ben Raskin

Figure 6.3: Silvopastoral system, cattle browsing trees at Eastbrook Farm.

Silvopastoral systems

These systems combine tree crops with the grazing of animals. Grazing sheep or cattle can only coexist with large trees, otherwise the animals will eat the tree crops. The sorts of trees associated with these systems are standard fruit trees such as apple, pear or olive, whose fruit can be knocked down and collected from the ground, and cork trees as found in Spain. Biomass trees, grown for the purpose of burning or chipping their timber, are grown in a pollarded form to keep the tree crops out of the way of the animals. Sometimes the trees chosen are also a fodder crop so they can be cut and fed to the animals, or browsed in the field. Sometimes tree crops are protected from grazing by fencing systems.

Poultry can be combined with smaller trees; chickens, geese, ducks and turkeys all fit well into agroforestry systems. Chickens are natural forest dwellers and thrive in a partially forested landscape. They feed on insects if they are available and many more of these will be found among rows of trees. Running poultry between coppiced trees or semi-dwarf fruit trees provides the birds with shelter and plenty of foraging opportunities.[6] The chickens also break the life cycle of fruit tree pests and provide fertility for the tree crop. Wolfe and Cindy Engels went one step further by providing herbs in the system so the poultry could self-medicate if they became ill.[7]

Silvoarable systems

These systems intercrop trees with arable crops (Figure 6.4). The trees are spaced to allow access for small tractors to manage fodder, grain, vegetable or fruit crops growing in the space between them. The trees in these systems can be smaller and lower to the ground, since they will not be damaged or grazed by animals.

Figure 6.4: Silvoarable, Wakelyns agroforestry.

Alley cropping tree crops uses

The tree and crop combination and design are key to the success of agroforestry systems. The choice of tree and crop combinations is a long-term decision and requires careful planning.

The trees provide a crop and environmental services just as do the horizontal crops or livestock that will be growing or grazing in between. The types of crops and services that trees can provide are listed below.

Common crops from trees (see Table 6.1)

- *Fruit.* These are the normal fruits such as apples and pears and in temperate or tropical climates could include olives or crops such as Sichuan peppers.
- *Nuts.* A wide range of nut crops on tall or small trees are grown around the world.
- *Nursery trees.* The trees are grown and then lifted and sold to a landscape company to be planted elsewhere.
- *Biomass for fuel.* Timber can be used as firewood and cooking fuel. Coppicing or pollarding systems provide biomass and the timber will regrow on a two to 10-year cycle, depending upon the tree. The thinner wood and brash from coppice can be chipped or made into pellets to be used in modern biomass boilers to heat the home or to heat glasshouses to grow out-of-season crops. The technology is being developed to provide heat and power from coppiced wood (combined heat and power [CHP] biomass boilers).
- *Biomass for soil fertility.* Ramial wood chips are made from the brash, or thin branches and twigs, harvested from a coppice system. Any branches less than seven cm thick are chipped and put directly on to the ground. They put high levels of carbon and nitrogen into the soil and increase the soil's organic matter, water-holding capacity, nutrient levels and microbial activity. Biochar can be made from this timber and used to increase soil fertility.
- *Livestock fodder.* Livestock have traditionally been fed with branches of trees as a winter feed. This is called 'tree hay'. Many trees offer medicinal benefits to animals. Hazel leaves have high tannin levels and will act as a dewormer for cows and sheep. Willow bark has high levels of salicylic acid (the natural form of aspirin) and animals will graze on it naturally if it is available, without killing the tree.
- *Building and craft materials.* In ancient systems pollarded oaks would provide building timber and today the use of timber is undergoing a revival. Trees continue to provide materials for crafts such as charcoal-making, basket-making, clog-making, thatching spars, pea and bean poles and hurdle- and fence-making.
- *Medicines.* Trees can be grown to provide medicines – for example, elderflower, elderberry and sea buckthorn. In tropical countries moringa and neem are often grown. Lime and hawthorn leaves can be made into teas. Birch sap is a highly valued drink. These are preventative medicines and can be processed and sold as high-value products.

Functions of trees (see Table 6.1)

- sequestration of carbon
- slowing down the movement of water through the soil
- offset of flooding – riparian buffers, tree-lined watercourses
- windbreaks

Tree	Growing Method	Use
Hazel	Coppice 5- to 10-year cycle	Biomass (five times the calorific value of willow) Thatching spars Small nut harvest Pea and bean sticks (substitute for bamboo imports) Hurdles CHP systems
Willow	Coppice biennial method Can also be pollarded	Biomass Basket-making Charcoal-making Hurdles CHP systems Ramial wood chips
Fruit trees: apple, pear, plum, cherry, Sichuan pepper	Semi-dwarfing or standard Annual harvest	Fruit harvest for eating, preserving, juicing, or fermenting into cider or perry
Nut trees: walnut, cob hazel, sweet chestnut	Standard varieties Annual harvest	Nuts
Small trees for livestock fodder: hazel, hawthorn	Semi-vigorous trees 2–3m tall	Cut or grazed as animal fodder
Wild harvest trees: elder, blackthorn, sea buckthorn, birch, lime, hawthorn	Annual wild harvest	Forage crop Sap, cordials, teas, medicinal uses
Standard form timber trees: ash, chestnut, poplar, walnut, birch	Single harvest when trees full grown; or coppiced on a long cycle.	Timber for building and furniture-making Pulp for paper Firewood
Standard trees for environmental services: alder	No harvest	Nitrogen-fixing environmental service

Table 6.1: Some of the trees that are commonly in use in agroforestry systems in Northern Europe and the US.

- pollinator fodder
- reduction of noise
- living snow fences
- living livestock fences
- reduction of smells from livestock operations
- habitat for biodiversity and functional biodiversity
- reducing energy requirements of buildings via cooling or windbreaks.
- fixation of nitrogen
- cooling down or warming up soils.[8]

Planting spaces and orientation in alley cropping systems

In larger-scale systems the trees are planted at a minimum of 12m, maximum of 28m, and at multiples of 4m to allow for tractor access and turning circles. Where possible the trees are planted on a north–south axis to allow the maximum light penetration into both the tree and the arable or vegetable crop underneath. This can be easily achieved on a flat site.

If the site has a slope, there are different considerations to take into account. Running the rows along the contour will reduce soil erosion and slow down water movement down the slope; it is easier to work by hand along the contour. For ease of management the tree rows need to be parallel. It is more challenging to operate a tractor or combine harvester on the camber, or across the slope, and this can be dangerous if the slope is steep. On a shallow slope the machinery can be adapted to the camber. On a steep slope it is preferable to align the rows up and down the hill, but this will compromise the trees' capacity to slow down the flow of water down the slope and so reduce the soil erosion that it causes.

Rows of trees running in an east–west direction will shade out the arable and vegetable crops to some extent, and if the tree crop is a fruit crop this will cause uneven ripening of the crop. A timber crop will not be affected by differential light levels. The tree rows will offer some form of wind protection, so they can be orientated to provide the optimum protection from the prevailing wind, although fruit tree crops will themselves need wind protection.

Each agroforestry design is customised to the soil type, the aspect and wind direction and the long-term use of the tree crops. The overall design of the tree row orientation should aim to provide the maximum environmental services and maximum yields while remaining within the management capacity of the farm and farmer.[9]

2. The forest garden or food forest

The forest garden is a traditional farming system found across the world. It is more challenging in the low light conditions of Northern Europe.

It was pioneered in the UK by Robert Hart in the 1980s and this in turn inspired Martin Crawford to plant up what is today one of the best examples of a forest garden in the UK.[10] Bill Mollison and David Holmgren favoured the forest garden as a system, especially in tropical agriculture conditions, and they incorporated it into the basic permaculture design course; it is thereby taught across the world and consequently there are hundreds of examples around the world.

The underlying concept of the forest garden is to mimic the 'climax vegetation' of a forest, that is, a stable, dynamic forest habitat, by substituting edible and other crops for the typical woodland trees. Once it is established, the whole system produces high yields of food and other products for very little work.

A forest garden is made of seven layers: canopy layer, small tree layer, shrub layer, herbaceous perennial layer, annual layer, root layer and climbing plant layer (Figure 6.5; Table 6.2).

The forest garden can be simplified, especially in lower light levels, by reducing the number of layers. Often a three-layered system is grown with a small tree layer, a shrub layer and a herbaceous perennial layer; this allows more light into the system and makes management much easier.

Forest gardens are mostly managed by hand and are not planted in straight rows. They are three-dimensional gardens full of perennial food plants. They can vary hugely in size; they can fit into small urban back gardens, or large community gardens, or form part of a permaculture farm or community enterprise.

Layer	Type of Plant	Use
Canopy layer	Walnut, sweet chestnut Standard apple or pear Medlar Mulberry Birch	Nut crop Fruit crop Fruit crop Fruit crop Sap
Small tree layer	Half-standard apple, pear, plum, cherry Hazel (filbert) Elderflower Sichuan pepper Coppice hazel, willow, sweet chestnut Crab apple, hawthorn, rowan	Fruit crop Nut crop Flowers and berries for cordials and health-giving properties Peppercorns Biomass for burning, charcoal-making, support sticks and fencing materials Fruits for jams and winemaking
Shrub layer	Gooseberry, blackcurrant, raspberry Rose *Eleagnus* spp.	Fruit crop Teas and syrups Nitrogen-fixing and berries
Herbaceous layer	Rosemary, thyme Strawberry, rhubarb Globe artichoke, perennial kale	Herbs and medicinal use Fruit Vegetable
Annual layer	Calendula New Zealand spinach Borage, nasturtium Salad greens	Edible flower and medicinal uses Edible greens Edible flowers, bee fodder Salads
Root layer	Jerusalem artichoke, yakon, wasabi	Edible roots
Climbing layer	Grapes, kiwi	Fruit crop

Table 6.2: Types of plants that are commonly used for the layers of a forest garden in Northern Europe or the US.

The food coming out of them can feed a family or a community and is highly prized by top end restaurants.

In tropical regions forest gardens are a common form of farming with complex polycultures of fruit, tree vegetables, medicinal plants and timber crops. The concept of the forest garden can be adapted and used in a temperate region as an orchard cropping system by planting alternate straight rows of fruit trees, soft fruit and herbaceous fruit crops. This constitutes in effect three layers of a forest garden but in straight rows. It increases the diversity but can still be managed with relative ease and small machinery.

3. Woodland pasture and parkland

Rackham suggests that in the UK the original 'wildwoods' looked like woodland pasture, that is, a savannah-type landscape comprising areas of grass with single or small clumps of trees grazed by large animals, not the dense forest often imagined (Figure 6.6). In modern times we are used to this concept in the African plains; the suggestion is that this is what the pre-enclosed landscape looked like in the UK.[11]

THE SEVEN LAYERS OF A FOREST GARDEN

Illustrated by Dilly Williams

1. Canopy

2. Low Tree Layer

7. Vertical Layer

3. Shrub Layer

4. Herbaceous

6. Soil Surface

5. Rhizophere

Figure 6.5: Seven layers of a forest garden.

On these 'commons' people would have had rights to use the land for grazing, collecting firewood, making charcoal and grazing pigs on acorns or beechmast. Woodland pasture has had a revival with the renewed interest in 'wilding' areas of European land; when large areas are rewilded, the landscape resembles a savannah.

Rewilding involves enclosing large areas of land and removing all of the internal fences. Into this system large resilient livestock are introduced – pigs, cattle, deer, ponies – and then the system is allowed to evolve naturally – a hands-off, process-led approach. The most well-known experiments of this kind are the Oostervarderplassen in the Netherlands and Knepp Castle in Sussex in the UK. Knepp Castle was a 1,400-hectare farm on relatively unproductive land, where farming did not produce a profit and was clearly harmful to the environment and watercourses. Charlie Burrel and his wife Isabella Tree took the unusual step of rewilding the entire estate. When the wild animals on it reach the carrying capacity of the land, that is, they run out of food, they are culled and sold as wild meat. The farm is now full of wildlife and biodiversity; many of the UK's rarest species turn up here in great abundance. The story is beautifully told by Tree in *Wilding*.[12]

It could be argued that this kind of landscape is similar to the 'parkland' that was created around large houses in the 18th century by landscapers such as Capability Brown. Large trees are dotted around the grazing areas surrounding wealthy landowners' homes, which are sometimes stocked with deer, cattle or sheep. These parks are much tidier than woodland pasture and were often created at the expense of the local people, who would be cleared from the landscape to make way for views for the estate owner.

4. Riparian buffers and shelter belts

Riparian buffers are groups of trees planted around watercourses. They prevent the erosion of soil into watercourses and cool down the water to provide a better environment for aquatic life. They also slow down the movement of water over the soil, helping water to penetrate into the soil by up to 60 per cent, as suggested in a case study in *The Agroforestry Handbook*.[13] This reduces flooding further downstream, since water is absorbed into the soil and tree roots rather than producing a fast runoff in severe weather.

Shelter belts are strips of trees and bushes planted to reduce wind speed. They can increase the temperature in a cropping area and reduce the evapotranspiration rate from a crop so that it needs less water. In fruit orchards they will allow the pollinating insects to do their work, so catalysing a higher yield. Shelter belts can be multifunctional and provide a wild harvest from trees such as birch and elder, a nutritional benefit to crops from alder, or timber from long-term coppiced timber trees.

Do agroforestry food production systems work?

So why go to the bother of planting thousands of trees in a farming system, at great cost in terms of the trees and labour, and with the potential consequence of making the management of the site more difficult? There are huge benefits in having a tree crop in a farmed landscape.

Increased yields

The trees provide an extra cropping yield in the long term. This yield is often of a high value in comparison with the arable crop or the grazing opportunities underneath the trees. It may be an annual harvest, or a one-off harvest after a number of years or a regular harvest at intervals of two to five years. The trees do not compete with the overall yield of the crop apart from the reduction of space that is given to the crop.

The total productivity of the integrated cropping is measured using the land equivalent ratio (LER). This is the ratio of the area under sole cropping to the area under intercropping needed to give equal amounts of yield at the same management level. It is the sum of the fractions of the intercropped yields divided by the sole crop yields. For instance:

To give an example quoted in *The Agroforestry Handbook*: with dual crops of walnut and wheat, you may achieve 40 per cent of the walnut yield and 80 per cent of the wheat yield, which gives a sum of 40/100 + 80/100 = 1.2. The LER in this case is 1.2, which means in effect that the combination of the two crops in the same space gives a total yield 20 per cent higher than the yield of just one crop. *The Agroforestry Handbook* suggests that most researched agroforestry trials show an overall increase of 20–40 per cent in total yields from a combined agroforestry system, without including the environmental services provided from these systems.[14]

$$LER = \frac{\textit{Yields of agroforestry tree crop} + \textit{Yield of agroforestry arable crops}}{\textit{Yield of solo tree crop (or Yield of solo arable crop)}}$$

Carbon sequestration

Carbon sequestration is increased in the total system because the trees soak up and store carbon in their wood. *The Agroforestry Handbook* quotes research showing carbon sequestration rate in the UK of one to four tonnes per hectare per year from agroforestry tree densities of 50–100 trees per hectare.[15] Jules Pretty estimates the rate to be 6.6 tonnes of carbon per hectare per year in a tropical agroforestry system in the climates of India and China.[16] Martin Crawford estimates carbon sequestration rates for a full forest garden type of agroforestry system in the UK to be 4.8 tonnes per hectare per year.[17] Even if these trees are harvested and burned as fuel, releasing their carbon again, it is better than burning fossil fuels, since this practice involves a short carbon cycle rather than releasing carbon that has been stored for millions of years as fossil fuels. If the brash from the coppice system is chipped and used as ramial chips the organic matter and fertility of the soil are increased further, storing more carbon and water.

Water management

Water management is improved by planting trees in the fields or across slopes, since this reduces the flow of water through the landscape, allowing water to penetrate deep into the soil. This reduces flooding and the erosion of soil during heavy rains and storms.

Crop nutrition

Trees mine nutrients deep from the subsoil, and when they drop their leaves these nutrients are released into the topsoil. The trees increase the organic matter, the biological activity and the

Ramial wood chips – by Christopher Upton, Zerodig

Ramial wood chips are made from woody material that has a stem diameter of seven cm or less. This material has a relatively high proportion of cambium, the green part of the twig just under the bark, which contains more minerals, sugars and complex plant compounds than the lignin, or woody part of the twig. This means it has a low carbon to nitrogen ratio and breaks down easily. Ramial wood chips are made in the winter when the trees are dormant and stored nutrients are at their highest level. Using ramial wood chips as the primary source of organic matter ensures a rich, healthy and fungal-dominant soil biology. When the soil food web includes a strong fungal network, plant growth goes into overdrive. Plant roots will connect with fungal hyphae, called mycorrhizae, which can amplify the reach of the plant roots at least 100 times, perhaps up to 1,000 times. This means the plant is more drought tolerant and can access nutrients from a much wider area. Fungi also synthesise many complex biological molecules that are important for both plants and our own health.[18]

mycorrhizae in the soil, all of which increase the soil's fertility and therefore crop nutrition. The mycorrhizae also form networks in the soil which distribute the nutrients to other plants and trees in the system. To quote Peter Wohlleben in *The Hidden Life of Trees*:

There are fungi present that act like intermediaries to guarantee quick dissemination of news. These fungi operate like fiber-optic Internet cables… the fungal connections transmit signals from one tree to the next, helping the trees exchange news about insects, drought and other dangers… This equalisation [of sugars] is taking place underground through the roots. There's obviously a lively exchange going on down there.[19]

Both people and animals love trees; trees improve the working environment for people and the grazing environment for animals by providing shade in the heat, and wind protection in the cold, and preventing snowdrifts in cold climates. This reduces the fuel requirements of houses or buildings surrounded by trees. For crops growing in between the trees the temperature is increased when it is cold, and decreased when it is hot, and the wind speeds are reduced so the crops need less water and hence less irrigation.[20]

Biodiversity

Biodiversity is increased in agroforestry systems. As the organic matter content of the soil is increased so are the numbers of worms, insects, beetles, small mammals and birds. The trees themselves offer roosts for birds, and if grasses in the understorey are left to grow long and produce seeds they will provide a rich source of food for seed-eating birds. Insect-eating birds will also find food in plentiful supply, and the predatory birds will snack on the small mammals in the tree rows. Butterflies and moths will find good habitat and food sources in agroforestry rows. Occasionally, this biodiversity causes problems in the crops, when voles eat seed crops or birds eat fruit, but once the ecology has become established the checks and balances normally prevent these problems.[21]

Less habitat fragmentation

Habitat fragmentation is a big contributory factor in the loss of biodiversity. With climate change, species need to be able to migrate to adapt to the food supply and weather conditions. With the fragmentation of habitats, it is difficult for the isolated populations to meet up to breed, or to find the next spot where there is a food source. The tree rows offer corridors for the movement of many species across the landscape.[22]

Reduced environmental stress

Trees in a landscape reduce environmental stress and, in turn, the stress that people experience. People enjoy a landscape with trees in it. Millions of tax pounds are spent in the UK preserving traditional landscapes with trees in them. Work within an agroforestry system is a much kinder psychological and physical experience than work in a conventional horticultural or arable field. Once trees have been integrated into the system, the landscape becomes more human scale, work comes in manageable parcels, trees provide shelter from the wind, and people are inside a landscape rather than on it. These ideas are backed up by research done by psychologists Kaplan and Ulrich. They have shown that as soon as people move into a 'park-like' landscape with trees in it their heart rate slows and their blood pressure goes down. There is a suggestion that human beings are hardwired to enjoy a savannah-type landscape from our deep past on the African plains. The view of a tree from a hospital bed increases recovery rates, and the view of a tree or even a picture of tree visible from a dentist's chair reduces the amount of pain relief required.[23]

Less soil compaction

Smaller machinery is needed with the introduction of trees, resulting in less soil compaction.

Challenges of agroforestry

Introducing trees into a landscape does make the farming system more complex and this brings challenges. It is rare to find all the skills required in one person. This problem can be overcome by using teams of people or nesting micro-businesses in the farm. The complexity and plant knowledge required to manage a full forest garden will be difficult for all but the most highly trained. A simple alley cropping system is much easier to manage. Designing the orientation of the rows and the choice of trees does require expert knowledge. The capital cost of the trees can be very high, but this can be covered by the extra income from the tree crops.

What does an agroforestry food production system look like?

Modern agroforestry systems are now reaching maturity and have been studied for some time. Experimental systems are now being replicated as sustainable cropping systems. A further case study of agroforestry can be found in Chapter 10, in the agroforestry design and implementation at Huxhams Cross Farm.

Case study: alley cropping – Wakelyns Farm, Fressingfield, Suffolk

Martin Wolfe and his wife Ann began to design and plant Wakelyns Farm in north Suffolk in 1994. It remains an ongoing experimental site and is one the most established alley cropping agroforestry systems in the UK. Martin and Ann designed and created this radical farm in their retirement, and it is truly a farm for the future and an oasis of biodiversity in the East Anglia plain of the UK.

The farm is on 24 hectares of heavy and flat clay soils in one of the driest areas of the UK, with 50 cm of rainfall per year. It has a more continental climate than much of the UK, with hot dry summers and wet chilly winters. In 2019, the temperature reached 38.7°C in nearby Cambridge, the hottest temperature ever recorded in the UK.

Wakelyns is set among the prairielands of East Anglia. To get there you drive through miles of the UK wheat belt, where the farms are often 400 hectares each, where most of the hedgerows have been removed and wheat is monocropped up to the edge of the road, not a weed in sight. This is efficient industrial farming, producing some of the highest yields in Europe. However, when you turn the corner into Wakelyns you meet a landscape full of trees and people and exploding with birds. Owls fly up and down the tree rows patrolling for mice and voles that have been feasting on worms and beetles and seed. Fieldfares regularly stay all year round, as monitored by the British Ornithological Trust.

The site was designed with four trial 'fields':

- alder for nitrogen-fixing, apple and cherry trees for fruit crops, and oak and lime as nursery stock and timber trees
- hazel coppice for biomass and hedging and craft uses
- willow, a fast-growing biomass variety
- walnut trees, plums and apples.

The rows are spaced apart 12m, 15m or 18m and run north–south (the newer tree lines were planted with a wider spacing). The site is flat, so the north–south orientation of the rows works well at Wakelyns, which is registered and run as an organic farm. In between the tree rows are grown vegetables (now managed as CSA using the 'no dig' model), YQ population wheat, lentils and other grains. A flock of chickens for eggs has at times been added into the rotation.

Maximum diversity has been squeezed in at all levels of the farm. Rather than grow one variety of lettuce, they will grow five. Trials with potatoes have shown that when three different

Figure 6.7:: Wakelyns agroforestry drone picture.

With kind permission of Martin Wolfe

varieties are grown together the spread of disease – potato blight – is slowed down by as much as 50 per cent. The blight trials on potatoes show that growing together red, white and knobbly potato varieties, chosen carefully so they share a harvest date but can be sorted out after harvest or sold as a mixed bag, reduces blight by 50 per cent without any sprays at all. The spread of disease is slowed because of the different genetic resistance found in the different plants. Martin and Ann developed these polycultures, as opposed to the monocultures prevalent in industrial farming systems, on many levels in the cropping systems, an approach that has continued even since their deaths in 2019 and 2016, respectively.

Wakelyns continues to demonstrate success in growing a wide diversity of crops, with agroforestry tree species mixed with arable and horticultural crops. The diversity is increased at a varietal level by growing multiple varieties of each crop and at a genetic level by choosing open-pollinated varieties or by recreating open-pollinated varieties of wheat that were not otherwise available.

The Wolfes realised that modern wheat varieties are not suited to a low-input organic agroforestry system. So, in conjunction with the Organic Research Centre, they created a new 'population wheat' by crossing 20 varieties of wheat found in industrial agriculture by the John Innes Institute. The modern varieties of wheat were chosen for their yield and quality, so the new population or landrace wheat is now called 'YQ wheat'. When YQ wheat is grown the genetic diversity is so great that the crop is very

Figure 6.8: Martin Wolfe at Wakelyns Farm.

resistant to disease and when grown on different sites across the UK and Europe it readily adapts to the new soil and climate. The farmer can then keep seed back for sowing the following year so that the wheat adapts to the particular site it is in. Population wheat commonly produces 70 per cent of the yield of an industrial variety, but without the inputs. YQ wheat has a very strong flavour and is naturally low in gluten. Martin and Ann bought a small mill and started to mill flour on their farm, selling it fresh and local.

When this wheat grows in between rows of trees that are themselves genetically diverse the system becomes even more rich. In partnership with Hodmedods, Martin and Ann trialled the growing of lentils, chickpeas and chia in the UK.

The tree crops are all used. The willow and hazel coppice fuel a biomass boiler to heat the house. Experimentation in a CHP system is planned, in which the boiler that heats the water for the heating systems will also generate electricity. Martin and Ann found that the apple crop has less scab because it is not near any other fruit trees, although the apples are difficult to pick because the trees are spaced out. The alder provides nitrogen. The trees initially intended for nursery stock have remained and are now being pollarded, and chipped for the production of ramial chips to increase soil fertility. Produce from all aspects of the farm is now also being used in the on-site bakery.

Produce from the farm is sold via Hodmedods and into the London markets and CSA such as Growing Communities, as well as more locally. Martin and Ann were pioneers in the decommodification and relocalisation of grains and pulses. They were never afraid of trying something out. Some ideas did not work; some did. Pioneering these systems takes perseverance and the ability to learn from doing, and to learn from things that don't work as well as those that do work.

Martin and Ann always had three pieces of advice: always plant trees; maximise diversity in your cropping; and there is no such thing as a failed experiment, only more data.[24] The Organic Research Centre has published much of the data from these experiments.[25] Following their deaths, their family is continuing the organic agroforestry and the traditions of innovation and research, while also developing Wakelyns into a hub for food, farming and biodiversity-related activities (Figure 6.7).[26]

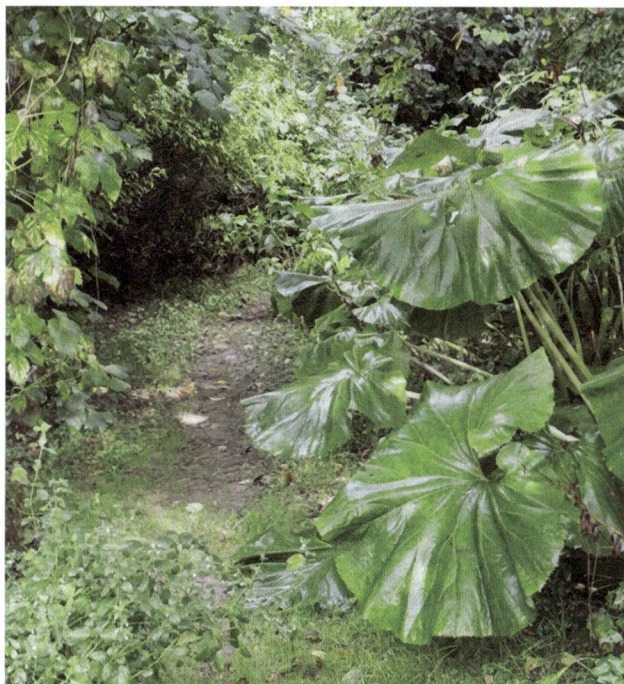

With kind permission of Walter Lewis

Figure 6.6: Agroforestry Research Trust forest garden.

Case study: forest garden system – Agroforestry Research Trust, Dartington, Devon

In 1994, Crawford developed the Agroforestry Research Trust and planted a forest garden on a 0.8-hectare site on the Dartington Hall Estate nestled next to Schumacher College near Totnes in South Devon, UK. Shortly after, he also planted a 3.3-hectare nuttery and unusual tree fruit trial site, and in 2011 he developed a 4.5-hectare tree nursery and glasshouse site a few miles away.

The forest garden is on clay soils in an area of high rainfall, 100 cm per year. It is relatively sheltered from the prevailing southwesterly winds, but can experience strong gales, especially in the winter months. Crawford suggests that the climate that is experienced in the southwest of England has become more like that of the northwest of France, so he is experimenting with future-proofing his forest garden with species and tree crops not normally grown in the area.

The Agroforestry Research Trust is a charitable trust, and the sites are partially experimental. Crawford has pioneered growing a huge range of perennial crops that are not normally cultivated in Europe for food, medicines or fibres – such as Sichuan peppers, toon tree vegetable and New Zealand flax. He has written several books on the growing and eating of a huge range of crops and has appeared on BBC Radio 4's *Food Programme*, among others, demonstrating his cuisine of the future. He offers online and in-person courses and also tours of the sites. Income is derived from plant sales, tours, training and book sales.

The main forest garden site is made up of the seven layers of canopy trees, smaller trees, shrubs, herbaceous perennials, climbers, annuals and mushrooms. It is one of the most mature sites in the UK that demonstrate this type of food production. Crawford's research suggests that the 0.8-hectare site alone sequesters four tonnes of carbon per year, equivalent to 4.8 tonnes of carbon per hectare per year. The SOM rose from eight per cent in 1994 to a huge 16 per cent in 2019; this means the soils are more resilient to warm dry summers and heavy winter rainfall.

The nuttery includes chestnuts, hazel and walnuts, but is also trialling butternuts from the US, black locusts from Hungary and persimmons.

The glasshouse, on the third site, is 500m^2, and Crawford is experimenting with a tropical forest garden of citrus, mango, avocado, papaya and banana. The glasshouse requires no fossil fuels but is heated from heat stores created by a 'wall of water' made from black IBCs (international bulk carriers) stacked three high and filled with water. At night these re-radiate heat that has been stored in the daytime. The soil also stores heat from the day and radiates it at night. The hot air is sucked into the soil via a system of pipes and tubes and a fan. The business is self-funding with some input from grants and donations.[27]

Case study: agroforestry in a commercial food production system – Broadlears Agroforestry, Dartington Hall, Devon

What is interesting about forest gardening is that it has many offshoots. Dartington Hall Trust recently planted up one of its prime 20-hectare arable fields as an alley cropping silvoarable system (Figure 6.9). The agroforestry rows are composed of three types of trees: elder grown by Luscombe Drinks, apples grown by the Apricot Centre (Huxhams Cross Farm), and Sichuan peppers – a new crop that has been pioneered, propagated and grown by Crawford just a short distance from this field for the London Peppercorn Company.

The tree rows are on 20m spacing and aligned north–south. In between the rows arable crops are grown for use by the Dartington Hall farmer Jon Perkin. These crops are YQ population wheat – a new landrace wheat, called Cornovii, bred in Somerset by Fred Price for the new Dartington Mill and Almond Thief bakery – together with fodder for the goat herd, and hemp that is made into oils, seeds, cosmetics, hempcrete and building products on a four-year rotation.

What is especially pioneering about this system is that it has five partners, all growing crops in Broadlears agroforestry field: the four companies growing their crops and Dartington Hall Trust. The trust pioneered a form of tenancy agreement between the five parties which covered rents, tenancy and organic status over a 15-year tenancy period.

This agreement was created because none of the companies or farmers involved had the complete skill set or equipment to run the agroforestry system in its entirety. This system has allowed access to land for the tree crop growers, in a cheap and simple way, and additional income for the main tenant farmer with only a tiny loss of cropping space. The arrangement is in its infancy, but the cropping space is certainly more diverse than one simply used to grow barley or maize as a fodder crop.[28]

Photo: Christian Kay

Figure 6.9: Broadlears Fields, Dartington Hall Estate, agroforestry rows.

Training in agroforestry

Some universities offer master's degrees in agroforestry, for example the University of Bangor in Wales. Training for farmers on introducing and designing agroforestry systems can be found around the globe. Or you can teach yourself using the resources listed below. Forest gardening is taught at an introductory level as part of all permaculture design courses and in detail at, for example, the Agroforestry Research Trust in the UK. Courses can also be found worldwide, especially via local permaculture groups.

Where is the agroforestry movement going?

Agroforestry is by its very nature a long-term farming system. It is of huge benefit in the development of modern sustainable farming systems that aim to tackle climate change and to integrate biodiversity into the heart of farms. It is still at the research and policy stage. The case studies highlighted here were both started in the mid-1990s on an experimental basis.

However, after 20 years, as the findings from these sites have become clear, more commercial agroforestry sites are being planted.

The recent 'Field Guide for the Future' from the Royal Society of Arts (RSA) Food and Farming and Countryside Commission in the UK recommends agroforestry as a way forward to transition to a more sustainable form of farming, with extra yields, more carbon sequestration and more biodiversity.[29] The FAO has also promoted agroforestry as a good method of sequestering carbon while providing fuel, food, fodder and medicine to people in the global south and improving water management.[30] ∎

In a nutshell

- Agroforestry is the integration of trees within a farming system.

- This may be in the form of alley cropping, that is, strips of trees within an arable system or within a pasture system.

- The tree crops may provide a food yield or an environmental service yield.

- Forest gardens are complex gardens made of edible trees, bushes and shrubs. These are found in traditional contexts around the world.

- There is no registered trademark for agroforestry systems.

- Agroforestry is frequently incorporated into other sustainable and regenerative farming systems.

Further reading, links and films

The Farm Woodland Forum is a charity and website disseminating information and running workshops on agroforestry in the UK and Ireland: www.agroforestry.ac.uk

Woodland Trust, 'The Role of Trees in Arable Farming', 2015.
Agricology: www.agricology.co.uk/field/farmer-profiles/stephen-briggs
Organic Research Centre UK: http://www.agforward.eu

The website www.agroforestry.org contains a huge amount of resources about agroforestry, mainly focused on farming in the Pacific and tropical regions.

P. Whitefield, *How to Make a Forest Garden*, Permanent, East Meon, UK, 2012.

The Agroforestry Research Trust website run by Martin Crawford has links to books, his courses and tours, and his tree nursery selling trees suitable for forest gardens in Northern Europe: www.agroforestry.co.uk

Film featuring Martin Wolfe: www.wakelyns.co.uk

Short films, virtual tours and podcasts from BBC Radio 4 may be found on www.agroforestry.co.uk

Films about the wilding at Knepp Castle and woodland pasture: https://knepp.co.uk

1 J. Giono, *The Man Who Planted Trees*, Chelsea Green, Hartford, VT, 1996.

2 M. Wolfe, 'Crop Strength through Diversity', *Nature* 406, 2000, pp. 681–682.

3 O. Rackham, *Woodlands*, William Collins, London, 2006.

4 http://www.fao.org/forestry/agroforestry/

5 Wolfe, 'Crop Strength through Diversity'.

6 S. Gabriel, *Silvopasture*, Chelsea Green, Hartford, VT, 2018.

7 C. Engel, *Wild Health*, Houghton-Mifflin, Boston, 2002.

8 https://www.fs.usda.gov/nac

9 Soil Association, *The Agroforestry Handbook*, Bristol, 2019.

10 M. Crawford, *Creating a Forest Garden*, Green Books, Dartington, UK, 2010.

11 Rackham, *Woodlands*.

12 I. Tree, *Wilding*, Picador, London, 2019.

13 Soil Association, *The Agroforestry Handbook*.

14 Ibid.

15 Ibid.

16 J. Pretty, *Agri-Culture*, Earthscan, London, 2002.

17 Crawford, *Creating a Forest Garden*.

18 https://zerodig.earth

19 P. Wohlleben, *The Hidden Life of Trees*, William Collins, London, 2015.

20 https://www.worldagroforestry.org

21 Soil Association, *The Agroforestry Handbook*.

22 Ibid.

23 R.S. Ulrich, 'Biophilia, Biophobia and Natural Landscape', in *The Biophilia Hypothesis*, ed. S. Kellert and E.O. Wilson, Island Press, Washington, 1993.

24 https://www.wakelyns.co.uk

25 J. Smith, 'Agroforestry: Reconciling Production with Protection of the Environment – a Synopsis of Research Literature', ORC, 2010.

26 https://www.waklyns.co.uk

27 Crawford, *Creating a Forest Garden*.

28 https://www.dartington.org/about/our-land/agroforestry; RSA, 'Our Future in the Land', 2019.

29 RSA Food and Farming and Countryside Commission, 'Field Guide for the Future', 2019.

30 http://www.fao.org/forestry/agroforestry/

Chapter 7

Agroecological food production systems

We are all bound by a covenant of reciprocity;
plant breath for animal breath, winter and
summer, predator and prey, grass and fire, night
and day, living and dying. Water knows this,
clouds know this. Soil and rocks know they are
dancing in a continuous giveaway of making,
unmaking, and making again the earth.[1]

R. Wall Kimmerer

Introduction

The 'green revolution', the term used to refer to the rolling out of industrial agribusiness across the global south, started in the 1960s and continues to this day. Replacing small-scale peasant farmers with large-scale crops for cash export, introducing pesticides, fertilisers, machinery, patented seeds and GMO crops, agribusiness is controlled by a few huge corporations. Forests have been cleared and 'peasants' pushed from their land to move into cities. Dams have been built for the purposes of hydropower and irrigation, displacing millions of traditional people. The green revolution did result in increased food production for a few years and did save lives in certain times and places. But the yields of monocrops are now in decline as soils become exhausted. The green revolution has destroyed traditional varieties

of crops, soils, forests, habitats, watercourses, balanced nutrition and local food cultures and livelihoods. We can learn from and do better than this with the same or less investment. The response to the green revolution in the lineage of sustainable agriculture is agroecology.

What is agroecology?

'Agroecology is a discipline that defines, classifies and studies agricultural systems from an ecological and socio-economic perspective, and applies ecological concepts and principles to the design and management of sustainable agroecosystems.' This definition is from Miguel Altieri and Clara Nicholls's handbook on agroecological practices.[2] A Chilean academic, now based at the University of California at Berkeley, Altieri is one of the founders and shapers of the contemporary practice of agroecology. 'Agroecology' has become an umbrella term used to describe both traditional and new farming systems that are based on a cyclical rather than a linear model of farming. This means that they involve creating 'closed loop systems'. Agroecology is a farming practice, a social and political movement and a scientific discipline found in universities across the world. It was first defined in the 1940s but really came to the fore in the late 1980s and the 1990s as Altieri and Stephen Gleismann gave it scientific and practical definition, holding conferences, publishing books and bringing together a community of practitioners and researchers.

Agroecology applies certain principles and practices not only to food production and growing but also to how the food is sold and distributed and how resources are managed by the local community. Agroecology is a farming system that includes the whole food system. It does not have a certification scheme. Farms self-define themselves as following agroecological practice; they may still use fertilisers and pesticides although in smaller amounts. They aim to create a closed loop system and to sell their food locally.

Agroecology has brought a greater level of political awareness to sustainable food systems as a whole. As a political movement it aims to steer the development of global food systems away from agribusiness and towards relocalised, healthy, sustainable and equitable food systems. It recognises that a systems rethink is necessary.

Agroecosystems managed according to these principles look very different from industrial agricultural systems, and are based on a different paradigm. Increasing the use of agroecological approaches in order to enhance the sustainability of food production would demand social and institutional changes in agricultural communities, the wider food system, and policies for agriculture, development and trade.[3]

La Via Campesina, or the International Peasant Farmers' Union, is the overarching organisation of the agroecology movement. Its international secretariat rotates through different countries and continents. It is a grassroots organisation made up of 182 local organisations. In the UK, La Via Campesina takes the form of the Landworkers' Alliance. In the US, it encompasses organisations such as the Rural Coalition. La Via Campesina is represented by organisations in 81 countries across the world. These organisations between them represent 200 million farmers. They define themselves as 'an autonomous, pluralist and multicultural movement, independent from any political, economic or other type of affiliation'. La Via Campesina has three aims: defending food sovereignty, promoting agroecology and the right to save local seeds, and promoting peasants' rights and access to land.[4]

Agroecology has become a political movement campaigning for peasants' rights to access to land, markets and resources. The concept of 'food sovereignty' was developed by the agroecology movement and is key in its campaigning. 'Food sovereignty' was defined in 2007 as:

…the right of peoples to healthy and culturally appropriate food produced through ecologically sound and sustainable methods, and their right to define their own food and agriculture systems. It puts the aspirations and needs of those who produce, distribute and consume food at the heart of food systems and policies rather than the demands of markets and corporations. It defends the interests and inclusion of the next generation. It offers a strategy to resist and dismantle the current corporate trade and food regime, and directions for food, farming, pastoral and fisheries systems determined by local

producers and users. Food sovereignty prioritises local and national economies and markets and empowers peasant and family farmer-driven agriculture, artisanal fishing, pastoralist-led grazing, and food production, distribution and consumption based on environmental, social and economic sustainability.[5]

Agroecology is the largest of all of the sustainable farming systems described in this book, in terms of land occupied and the number of farmers involved. There are no exact figures of how much land these farmers occupy. It could be six per cent of worldwide agricultural land, to judge by a recent study.[6]

Where did it start?

Agroecology started out life as an academic study of plant and crop ecologies in Germany and the US. However, it came to the fore in the 1980s and 1990s and became established across the world among groups of small-scale farmers. They recognised traditional practices of farming, combined these with new techniques and developed methods of sharing their knowledge.

As the green revolution has been rolled out across the global south many indigenous farmers have been displaced. Across Africa, Asia and South America 'peasants' and tribal peoples who have farmed land for centuries, but don't have legal deeds to their traditional homelands, have found themselves pushed from their farms. Millions have lost access to their land because it has been bought up by large companies for large-scale industrial farming.

Farms have been developed from cleared forest. Valleys have been dammed and flooded to capture water for hydroelectric power generation and to store water for the irrigation of land further downstream. Crops such as soya are grown on cleared Amazon rainforest land for export as animal feed for the cattle lots in the US and China. Palm oil, now ubiquitous in most processed food, has been planted on cleared rainforest land in Indonesia.

The human cost of this has been written about by Arundhati Roy, the acclaimed Booker Prize novelist. The Narmada Dam in India has displaced thousands of people from their lands.

In the winter of 1961 the tribespeople of Kothie, a small hamlet in the western state of Gujarat, were chased off their ancestral lands as though they were intruders. Kothie quickly turned into Kevadiya Colony, a grim concrete homestead for the government engineers and bureaucrats who would, over the next few decades, build the gigantic 138.68 metre-high Sardar Sarovar Dam… The people of Kothie joined the hundreds of thousands of others whose lands and homes would be submerged – farmers, farmworkers, and fisherfolk in the plains, ancient indigenous tribes people in the hills – to fight against what they saw as wanton destruction. Destruction, not just of themselves and their communities, but of soil, water, forests, fish, and wildlife – a whole ecosystem, an entire riparian civilisation. The material welfare of human beings was never their only concern.[7]

The agroecology movement developed into a political movement and in 1993 La Via Campesina, which literally means 'The Peasants' Way', was set up and is now arguably one of the largest social organisations in the world. Its members, small-scale farmers and food producers from around the world, deliberately chose the word 'peasant' to describe themselves. In each language this

Unity among peasants, landless, women farmers and rural youth

La Via Campesina is an international movement that brings together millions of peasants, small and medium-sized farmers, landless people, women farmers, indigenous people, migrants and agricultural workers from around the world. It defends small-scale sustainable agriculture as a way to promote social justice and dignity. It strongly opposes corporate-driven agriculture and transnational companies that are destroying people and nature.

La Via Campesina comprises about 182 local and national organisations in 81 countries in Africa, Asia, Europe and the Americas. Altogether, it represents about 200 million farmers. It is an autonomous, pluralist and multicultural movement, independent of any political, economic or other type of affiliation.

A group of farmers' representatives – women and men – from the four continents founded La Via Campesina in 1993 in Mons, Belgium. At that time, agricultural policies and agribusiness were becoming globalised and small-scale farmers needed to develop and struggle for a common vision. Small-scale farmers' organisations also wanted to have their voice heard and to participate directly in the decisions that were affecting their lives.

La Via Campesina is built on a strong sense of unity and solidarity among small- and medium-scale agricultural producers from the north and the south. The main goal of the movement is to realise food sovereignty and stop the destructive neoliberal process. It is based on the conviction that

small-scale farmers, including peasant fisher-folk, pastoralists and indigenous people, who make up almost half the world's people, are capable of producing food for their communities and feeding the world in a sustainable and healthy way.

Women play a crucial role in La Via Campesina's work. According to the FAO, women produce 70 per cent of the food on earth but they are marginalised and oppressed by neoliberalism and patriarchy. The movement defends women's rights and gender equality at all levels. It struggles against all forms of violence against women.[8]

Subtle agroecologies – by Julia Wright

The fact that our farming and food systems require a systemic change away from industrial approaches and towards greater ecological and social sustainability is only half of the challenge. In contrast to the industrial worldview, which perceives the need to accumulate material security and to hold dominion over nature, the worldviews of indigenous cultures recognise not only our interconnected relationship with nature but also the hidden invisible dimensions of life. These dimensions may be variously interpreted as involving vibrational energy, consciousness, spirit or something other-than-human. Whereas the majority of the predominant, ecologically based farming movements also remain rooted in the physical–material dimension, biodynamic farming embraces the invisible.

'Subtle agroecologies' is proposed as an antidote to this predicament. Rather than being a farming system in itself, it superimposes a non-material dimension upon existing, materially based agroecological farming systems. Grounded in the lived experiences of humans working on, and with, the land over thousands of years to the present it proposes that Western society needs a more appropriate consciousness, one that is ecological and grounded in the real and intimate connection between human and nature. In this sense we may conceive of the re-enchanting of agriculture as a way for modernist societies to reclaim their indigenous relationship with the living landscape they are in; a real-time, place-based relationship which may, therefore, be accessed and rekindled by anyone, anywhere. The following is a collection of techniques,

methods, arts and sciences associated with subtle agroecologies. This collection is not exhaustive, and many of the terms share similarities and may be used simultaneously: agro-homeopathy, astrology, biodynamic preparations, bio-electromagnetism, dowsing, eco-alchemy, feng shui/geomancy, interspecies communication, intuition/direct knowing, mantras/chanting, paramagnetism, planting calendars, prayer/intention, radionics, ritual, sacred geometry, sound/ultrasound, teacher plants/psychoactives, water dynamization.

Following this, the science or research of subtle agroecologies is the systematic study of the nature of the invisible world as it relates to the practice of agriculture. Depending on the situation, this research may take a goal-oriented, reductionist focus on, for example, increasing crop and livestock yields or reducing the incidence of pests and diseases, or a wide-angled vision of simultaneously working with multiple factors and concerns, all based on an ethics of care and with the overall purpose of bringing and maintaining balance and harmony to the farm, the farmer, the community and the world.

By working on vibrational, energetic dimensions, by becoming more adept at embodied practices that enable more conscious interaction with nature, and by re-evaluating our understanding of our place in the world, a systemic transformation of ourselves as human beings and therefore of agriculture is possible. This may go a long way towards achieving the goals that contemporary, ecologically based farming movements are ultimately aiming for.[10]

word has different connotations. The Spanish word 'campesina' means 'peasant' and is widely used in South America. If you look up the word 'peasant' in an English dictionary the definition is of 'someone of a rural labouring class' or 'someone who is uneducated and rude'. La Via Campesina has reclaimed the word to mean 'people of the land', people who live autonomous rural lifestyles and reject industrial farming methods.

Agroecology is also a social and economic movement. Its practitioners have tweaked and developed and shared traditional methods and new practices so that they can produce more healthy food but also support local ecosystems. It is about food and farming but also pioneers participatory or peer-to-peer learning among farmers; thus its methods have a strong social context. Agroecology works to develop local markets and trading systems, almost as an ecosystem in their own right. Having the right and ability to create seed-saving and seed-sharing systems helps small-scale farms to thrive through working in collaboration.

The agroecology movement has demanded and gained access to the decision-making processes at the UN, ensuring that the rights of millions of farmers across the world are respected.[9]

Agroecology is found all over the world, in both the global north and global south. It has a stated aim of recognising and valuing other world views besides the Western scientific viewpoint. However, as it is taught in universities across the world, it is the Western scientific viewpoint that often prevails. Julia Wright of the Centre for Agroecology and Water Resources at Coventry University explores these other world views.

Principles of agroecology food production systems

Agroecology has at its core a number of principles that are themselves based on the principles of ecology. These primarily aim to 'maintain ecosystem processes that perform central functions rather than performing them with substituted inputs, mechanisation and labour'.[11] These ecological principles are:

- Improve soil health.
- Conserve the efficient use of water.
- Sustain and improve functional biodiversity to create stable resilient and productive agroecosystems.
- Protect and enhance soil carbon and carbon sequestration.
- Create an equitable approach to the social and economic aspects of food production.
- Develop local marketing systems for both input and outputs from agroecology farming systems.
- Support smaller farm sizes to promote food security.
- Support shorter food supply chains.

Practices of agroecology food production systems

The practices that have been developed out of the principles are site specific. Some of them are itemised below alongside the principles they arise from:

- Improve soil health. Closing nutrient loops by growing green manures, especially legumes and catch crops, using composts and minimum tillage and mulching systems. Increasing microbial activity by the reduction of the use of pesticides. Reintegration of livestock on the farm, thereby increasing the amount of composting and organic matter returning to the soil.

- Conservation and efficient use of water. In practice farmers put in rainwater-harvesting systems for irrigation water, use drip irrigation systems to maximise the use of the water, mulch crops to reduce water loss, and plant agroforestry trees to reduce wind speed to reduce the drying of crops and the loss of soil through wind erosion.

- Sustain and improve functional biodiversity to create stable resilient and productive agroecosystems. An increase of agrobiodiversity within the cropping systems will preserve and create predatory habitats on the farm, allowing natural pest control to happen. The choice of crops that are genetically resistant to pests and disease, or so genetically diverse as to withstand pest and disease attack, coupled with the careful timing of their production reduces pest damage to crops.

- Protect and enhance soil carbon sequestration. In practice this includes increasing the SOM and microbial activity, planting agroforestry trees, and the use of minimum tillage systems, all of which contribute to climate change mitigation and adaptation.

- Create an equitable approach to the social and economic aspects of food production In practice, agroecology systems draw on the social context in their design, recognising that the solution to a problem is often found within local knowledge. Through including peer-to-peer learning in the process of creating these systems, the systems become unique to a given region. In the global south inappropriate systems are not imposed by well-meaning Western agricultural scientists; respect is shown to indigenous knowledge systems. In practice this includes using local materials, local expertise and local cultural knowledge.

- Develop local marketing systems for both input and outputs from agroecology farming systems. In practice, this increases food sovereignty. The food produced in agroecological systems is for local markets and home use rather than being grown as cash crops for export or supermarkets. Agroecology farmers do not need to buy in expensive seeds, fertilisers, pesticides or machinery. They create low-input and high-output resilient systems that provide food for local people even when growing conditions are challenging because of climate change. As many of these farmers are women, and as women tend to be the main food providers in the home, the children get an improved diet. Women often also incorporate simple medicinal plants into their farming systems.

- Support smaller farm sizes to promote food security. Smaller farms are often, arguably, more productive than larger industrial farms. They create regional food security and often work collaboratively in the form of cooperatives and other less formal networks to support each other and make the farming system resilient.

- Support shorter food supply chains. Shorter food chains between farms and consumers, as in CSA, mean that customers eat a more varied diet and less processed food and the farmers earn more money.[12]

What is the difference between agroecology and biodynamic, organic and permaculture systems?

Agroecology, biodynamic, organic and permaculture systems share many principles and practices. 'Agroecology' frequently acts as an umbrella term and it can encompass all of the types of farming practice outlined in this book. A biodynamic farm can be called 'agroecological', but an agroecological farm may not necessarily use biodynamic practices. A permaculture farm can be called 'agroecological'. but an agroecological farm may not use the permaculture design process.

Each of the sustainable food systems brings different attributes to the movement as a whole. Biodynamic farming brings a spiritual aspect, organic farming brings codification and standards, and permaculture brings design and social activism. Agroegolcogy has brought political awareness to the fore.

One of the principles of agroecology is to value different world views, and not to let the Western-trained scientific world view dominate. It is a challenge for scientists and practitioners, even agroecological ones, to value and understand these different world views. Even when agroecology is taught in universities, it is the science-based secular world view dominant in the global north that is normally presented. The biodynamic world view is one imbued with spirit, in which farming is linked to the rhythms of the seasons, moon and constellations. As such, it sits very comfortably as a European form of agroecology. Steiner possibly foreshadowed agroecology in his lectures when he said, 'Deciding to work together like this (mixing science and peasant wisdom) will be solidly conservative and yet also extremely radical and progressive … I hope this conference can become a starting point for genuine peasant wisdom to enter into the methods of science, which have become perhaps not stupid – that might be too insulting – but have indeed become dead'.[13]

As to differences, biodynamic and organic systems have the option to follow strict 'rules' that give them a certificate or standard that means their products can be sold at a premium price. They are codified. Agroecology and permaculture farms do not have a certification process.

Organic farming is sometimes equated with agroecology systems, but there are two major differences. Organic farming certification does not insist on the aim of closed loop systems and, although many organic farmers do work toward this aim, it is not required. An organic farmer is permitted to: buy in soil nutrients, but from a biological source; buy in the equivalent of pesticides, but ones that are more benign; and sell into a distant supermarket, flying products across the world wrapped in plastic. As the organic movement has moved into its third phase of development it has reflected on the fact that agroecology has had a much wider uptake than organic farming. This is partially because there is no need for expensive accreditation and certification in order to be an agroecological farmer.

Agroecology has a large uptake in the global south, where the movement is active on a social and political level as well. The lack of formal accreditation has made it relatively easy, cheap and effective for millions of farmers to learn and to implement.

Diversity and cropping systems

In enhancing the diversity of species in time and space the farmer creates complex polycultures of crops and introduces crop rotations. As a system this helps manage pests and disease and increases the resilience of the farm system in times of stress.

Increasing diversity is a key component of agroecology; an increase in biodiversity and agrobiodiversity makes a system more resilient. An increase in biological complexity can lead to an increase in the emergent properties of ecological systems. As the agroecological farm becomes diverse and more established it starts to function more effectively, often in unexpected ways. Increasing the genetic diversity in a crop or across crops slows down the spread of disease because the pathogen meets non-host plants and physical barriers. The physical barriers may be as simple as a wheat crop having variable lengths of stalk, so the ears are at different heights and a fungal spore splashing from one to the next will meet the stalk rather than the ear. 'Polycultures' are the polar opposite of 'monocultures'. A polyculture could involve growing an open-pollinated population wheat crop, or it might be a complex agroforestry system with three-dimensional stacking as found in a tropical forest garden. It might be simply growing an orchard with 10 varieties of apples rather than one, or it might be a fruit garden with 50 types of fruit with more than 100 different varieties spreading the cropping season. All of these scenarios make the farm more resilient and spread the risk to the farmer and help to ensure that they obtain a harvest or yield in all weathers and circumstances.

Crop rotations will in time reduce the build-up of pests, disease and the weed burden in the soil as well as balancing the nutrient draw on the soil and reducing the need for bought-in inputs to the system.

Increasing the overall diversity on an agroecological farm increases the 'functional biodiversity' that allows areas of wildlife or trees to become reservoirs of predatory insects and birds that then control the pests in the crops. This term refers to the work that biodiversity does for the farmer, reducing the need for expensive inputs of pesticides. Weed control is carried out by cultivations, either by hand or mechanically, mulching to cover the soil and so prevent weed germination, using fast-growing green manures to outcompete the weeds.[14]

Do agroecology food production systems work?

It has clearly been shown that agroecology systems work. They do what they set out to do, which is to 'maintain ecosystem processes that perform central functions rather than performing them with substituted inputs, mechanisation and labour'.[15]

A large-scale study in 2011 by Jules Pretty showed that agroecology farming methods can increase yields by 113 per cent. Forty initiatives across 20 countries were examined that employ agroecological production methods. These initiatives covered 12.6 million hectares and involved 10.4 million farmers. They included agroecological approaches to aquaculture, livestock and agroforestry, conservation agriculture and crop variety improvements with locally appropriate cultivars and cropping systems. Analysis of outcomes demonstrated not only an average doubling of crop yields, but also numerous environmental benefits, including carbon sequestration and reductions in pesticide use and soil erosion.[16] Follow-up research in 2020 showed similar results, with an increase in the number of groups of farmers and and the amount of land being farmed using these techniques, to over 8 million groups of food producers occupying 300 million hectares.[17] This amounts, arguably, to six per cent of the farmed land in the world.

Increases in yields and food security through agroecology improve nutrition, reduce rural poverty, contribute to climate change mitigation and adaptation and empower small-scale farmers across the globe.[18]

The ActionAid report 'Fed Up' gives an overview of how agroecology is so effective in achieving the multiple goals of increased yields, food security and climate change mitigation, and how this worked in many countries across the world. In Cambodia 65 per cent of rural farmers work very small plots of land, less than 1.3 hectares, and are often unable to feed their families, which causes them to reach for fertilisers to increase yields. However the Centre d'Études et de Développement Agricole Cambodgien (CEDAC) introduced the 'Multi-purpose Farm through Farmer Association' (MPF-FA) initiative in 2003. The specific method in Cambodia is to adopt the 'system of rice intensification' (SRI). Using known varieties of local high-yielding rice, the Cambodian farmers were encouraged to plant smaller areas of rice, feeding it with organic composted inputs and using natural pesticides and hand-weeding methods. In other areas of the small farms ponds were dug and used to store water and fish, frogs and ducks. The soil from the ponds was used for bunds, creating areas on

which to plant fruit trees and vegetables and to keep chickens and pigs. Not only do rice, duck, pork, vegetables and fruit make up the traditional Cambodian cuisine, but these farming methods allow families to feed themselves, sell the surplus on local markets and increase yields of rice by 60 per cent. These methods have been shared across Cambodia by the Farmer Schools – peer-to-peer learning groups that share best practice – and now 50,000 farmers practise similar methods.[19]

What does an agroecology food production system look like?

The farm will probably be small and employ many people, supplying local food to local people. It will use 'functional biodiversity' for pest control. The farms have diverse polycultures of plants and animals, with up to 100 species of plants found in one farm. They often include agroforestry and the use of legumes for fertility. The risk of crop failure is minimised by the stabilisation of diversity and yield over a long period of time. Agroecology farms are very resilient systems. They provide food security and produce food that fits the regional diet and cuisine. The farm will be socially connected, celebrating local cultures and festivals. It may reflect a spiritual approach to land, community food and farming.

Case study:
Green Acres Farm, Shropshire

Green Acres Farm is a 182-hectare certified organic farm in Shropshire. It is a mixed farm on which Mark Lea and his wife grow grains and pulses and graze cattle and sheep (Figure 7.1). The farm system is built on the belief that sustainability requires diversity.

The rotation is mixed with clover leys grazed by cattle and sheep as well as producing red clover seed. Crops include milling oats, pulses for human consumption grown for Hodmedods and 14 different milling wheats for direct sale to millers and bakers. The farm started working with Hodmedods in 2014. Lea valued peas in the rotation but did not feel they were fairly valued by markets, especially given how difficult they are to grow. They started growing yellow, blue, marrow fat and carlin peas for Hodmedods. This has not only helped them to maintain a diverse rotation but also helped them build meaningful connections with their customers. The carlin pea they grow for Hodmedods is known as Black Badger and is traditionally eaten on Bonfire Night (5 November) in the North of England. Companion cropping, diverse cover crops and agroforestry all contribute to the resilience of the farm.

Lea grows heritage varieties and population grain crops, which he thinks have a lot of potential in a low-input sustainable farming system.

Weed suppression can be fantastic; heritage and population wheats are good at scavenging nutrients and owing to their genetic diversity they have a resilience, to pests, diseases and climatic variability, that has been lost in the breeding of modern varieties.

Lea is building agroforestry into the farm system and is in the second phase of planting trees, in conjunction with the Woodland Trust. They have planted wide rows, with walnuts interspersed with hazels. They plan to cut the hazel on a five-year rotation, with 20 per cent cut each year, to provide wood for the biomass

With kind permission of Mark Lea

Figure 7.1: Green Acres Farm, making compost.

boiler. The rows run east–west to provide maximum shade and shelter on the north side of the agroforestry rows, where there is a field with a grass ley in rotational grazing. The concept is that the agroforestry rows will provide shade for cattle in summer and shelter for sheep in winter.

The farm is home to a green-waste composting enterprise that receives 5,000 tonnes per annum of local kerbside garden waste which is composted on the farm. The compost is used throughout the rotation on the farm to increase the SOM and improve the soil. The farm has been organic for 20 years and has proved that its system is sustainable without synthetic inputs. The farmers are still learning all the time. They are proud to farm in a way that contributes positively to biodiversity and soil, air and water quality while producing healthy food that is in genuine demand. The principles of agroecology are evident throughout the business. The way everything fits together has always been a priority and is essential in organic farming. Lea says,

It is all about diversity! First of all, it makes life more interesting! But we also think that diversity across the farm builds resilience. We are keen to increase diversity of the crops, the business and the people we work with.

Connecting with the people who eat and process our food makes us feel like food producers rather than commodity producers and is much more rewarding! Get closer to your market and become a food producer – bakers, millers – they will inspire you, give you new ideas. Don't be afraid to try things out![20]

Case study:
The Margarini Children's Centre and Organic Demonstration Farm – by Emmanuel Baya

The Magarini Children's Centre is a community-based organisation close to the Kenyan coast. Its main goal is to provide care and education to orphaned, vulnerable and marginalised children from the community, equipping them with life skills based on growing safe and healthy food using organic farming methods (Figure 7.2). The organic methods take care of the environment and the soil at the same time as giving the children a sustainable life and livelihood. The farm is the main source of learning for both the children and the community. The community meets every Thursday to learn organic farming methods. The method of learning is 'learning by doing'. As they do, they learn, and then they use what they've learnt back on their own farms. The crops that we grow are corn, cowpeas and beans. These are the staple foods of the children. We also grow vegetables like kale, sweet peppers, tomatoes, okra, mchicha and bringles.

We grow these crops by preparing the land by tilling it, to make it soft, and then making rows and adding our main organic fertiliser – 'bokashi compost' (a form of fermented compost). We plant the corn, cowpeas and beans directly in rows. The vegetables we raise in a nursery and then we transplant them.

We also make an organic pesticide using trees like the neem tree and other trees from the forest. We use crop rotation and we plant nitrogen-giving plants (legumes) to care for the soil. We mulch using dry grass. This is amazingly effective, since it rots back into the soil.

We make compost from the pigs', goats' and chicken's manure, in two ways. The first one is mainly based on the pig manure. The other one uses what's left from harvesting the corn and

Figure 7.2: Margarini Children's Centre and organic demonstration farm Kenya.

With kind permission of Emmanuel Baya

other main crops. We cover all of it with soil and let it decay, then we turn it after three months.

At the centre we have pigs, goats, cows and chickens. We treat these livestock as our partners in organic farming, since they provide us with manure that we use to nourish the soil. We also use permaculture skills and methods. We grow fruit trees. We have established a food forest at the demonstration farm. In the food forest we planted bananas, sugar cane, pawpaws, mangos and avocados in swales to collect water and prevent soil erosion by water.

We plant medicine trees such as moringa kungumanga (drumstick tree) and aloe vera.

The children have these natural medicines. They learn how to grow them and use them. Planting trees to make forest is one of our main events every year: we invite the community, and we plant trees at the centre and in the people's farms and by doing this we create awareness of climate change and show the importance of growing trees and not cutting them down

I also wish to share a success story about solving conflicts in the community and building peace using the soil. There has been a long conflict between the pastoralists and the crop farms. These pastoralists have large herds of cattle and are often nomads moving from one place to another. They really value their culture and have nothing other than their livestock and their culture to sustain their livelihood.

The crop farmers meanwhile value their crops very much to sustain their livelihood. However, when the pastoralists come to the communities with crops the goats destroy the crops. The farmers are Giriama people and the pastoralists are Gala people. The farmers would be so angry

that they wanted to kill the pastoralists and the pastorals would be very angry too. They were ready to kill the farmers. The government and non-governmental organisation tried to solve the conflict but it was escalating day by day and there was fear on both sides. Being from the Giriama community and a farmer, I wanted to see a sustainable solution to this conflict. It took me many days to figure out how to find an entry point into both communities. One day I went to the pastoralists. I wanted to know how they lived and also how they got their food, and more about their culture, so I asked the elders who are their leaders and decision-makers. I found out that they sell their livestock to buy food and also that they use the livestock as their food. I found out they have a culture of singing and dancing during the evenings. I was happy to discover that there were things that these two communities had in common, especially singing and dancing. The pastoralists speak the Gala language and a little Swahili and the farmers use the Giriama language and a little Swahili, so I was able to communicate in Swahili.

I told the pastoralists how good they were at singing and dancing and that I would like to create a dancing and singing exhibition for both the Giriama farmers and the pastoralists and they agreed. I asked the pastoralists if I could teach them Giriama singing and if they would then sing in their own language, and they said yes to my request. So I taught them to sing a song that values the cattle and that we are all from the soil. Both of our communities make the soil healthy to grow grass to graze the cows. The cows give meat and milk. For the pastoralists these cows are their source of food and livelihood. So the pastoralists composed some words as a song in their language.

I went back to the Giriama people and I asked them also to create a song that would value their food and crops like corn and all cereals and vegetables that are their life and livelihood.

This process took almost a month and a half. I asked the song composers from both communities, the pastoralists and the farmers, to teach their songs in their language to each other. The Giriama taught their song to the pastoralists and the pastoralists taught their song to the farmers, so the farmers were able to sing in Gala and in Swahili and the pastoralists song composers were able to sing the song in Giriama. When this was done I mobilised both the pastoralists' elders and the farmers' elders and then women and youth for a shared event of singing and dancing.

When the Giriama sang a song of how cattle are very important to the lives of the pastoralists, the pastoralists were very touched. The pastoralists sang the song of the farmers expressing how important the crops are to the life of the farmers, and the farmers were equally touched. They even shed tears.

After the sharing, the elders from both sides came together and made peace. The pastoralists promised not to bring their cattle to the farms, and the farmers also promised not to harm the cattle. They created a mutual agreement and the conflict was solved.

At the Magarini Children's Centre and Organic Demonstration Farm we have 286 boys and girls, 17 teachers, four security staff, four farmers, three cooks, one poultry attendant and two livestock staff. In total we have 28 staff who get a salary on a monthly basis.

To support a child to get education, food, stationery and books for one year is 35,000 Kenyan shillings (£233, US$323) and to pay a teacher for one year is 240,000 Kenyan shillings (£1600, US$2200). For the last six years water has been a big problem at the centre. Three years ago we received support to drill a 200 m deep borehole and we now have enough water. However, a water tank will enable us to collect rainwater as well.

What I have learned is that we all need and depend on each other. Together we can create a healthy and peaceful life.

Baya learned organic farming in Japan, and he is studying process-orientated psychology to get a better understanding of conflict resolution. He says that this farm is his definition of love. To sponsor a child or teacher or to help through one-off support of rainwater harvesting please contact Emmanuel Baya directly.[21]

Training in agroecology farming methods

In the global south Farmer Field Schools (FFS) are run to train farmers in agroecology methods. These schools are about both training and sharing knowledge to create unique solutions to local problems. Similar systems, in which farmers train other farmers, are operated in Europe and the US by organisations such as the Soil Association in the UK.

In Europe and the US, degrees in sustainable farming are generally offered in the guise of agroecology rather than organic or biodynamic farming. Modules on the latter methods may be offered under the agroecology umbrella.

Where is the agroecology movement going?

Agroecology is a science, a practice and a social and political movement and it is this combination that has made it the most successful of all the systems described in this book in terms of numbers of farms that have taken it up, numbers of hectares farmed and political recognition and support.

The FAO believes that agroecological farming systems are multifunctional and have the ability to solve many multifaceted problems and will help achieve many of the Sustainable Development Goals (SDGs). The FAO and UN support agroecology at a policy level and are working on methods to scale up the uptake of this farming method across the globe. The FAO suggests that agroecology offers a solution to feeding those without enough food in the global south, 75 per cent of whom are farmers themselves. If these farmers are helped to transition to agroecology methods they can increase their food production, become more resilient in the face of climate change and thus be able to feed themselves and their families. Agroecology will restore biodiversity, and it will sequester carbon, cut carbon emissions, ease pollution and repair damaged soils. The UN has now recognised La Via Campesina and given them a voice in food and agricultural debates. This voice is listened to by institutions such as the UN Human Rights Council and is broadly recognised among other social movements from local to global levels. It helps to balance out the power of multinational agribusiness in political forums.[22]

In the UK the Landworkers' Alliance has written a 'Framework for British Agriculture' policy document for UK agriculture after the UK's departure from Europe and the Common Agricultural Policy (CAP). To put this in context, the last time an agricultural bill was passed into law in the UK was 1947, and this ushered in industrial farming. The Agricultural Act of 2020 ushers in subsidies to pay for the provision of 'public goods' such as sequestering carbon and supporting biodiversity through farm practice. The Landworkers' Alliance has played a key role in lobbying, piloting and contributing to this new legislation. It has suggested policies to support agroecological farming, such as the relocalisation of food production, better access to land for food production, the right to live on small pieces of land, access to financial support, and subsidies for food production on a small scale. It has worked with the UK government to help shape a new agricultural policy that will include agroecology. Embedded within this new policy is the principle that subsidies are to be paid to farmers only if they provide 'public goods'. These public goods are carbon sequestration, clean water, clean air and supporting biodiversity, and all this is what agroecology farmers are good at.[23] ■

In a nutshell

- Agroecology is a farming and growing system that is built on the concept of the 'rule of return', creating farms with closed loop systems as much as possible.

- Plant nutrition is provided by composting plant and animal manures.

- Pest and disease control is through natural predators or cultural methods.

- Food sovereignty is a key concept: food is produced specifically for local sale.

- Farm crops are often integrated with livestock.

- Agroecological farmers across the world often have a world view in which plants and land are imbued with spirit.

- There is no formal trademark or registration system.

- Agroecology is one of the world's largest social and political movements.

Further reading, links and films

The Land is the quarterly magazine published by the Landworkers' Alliance: www.thelandmagazine.org.uk

Coventry University Centre for Agroecology and Water and Resilience: www.coventry.ac.uk/research/areas-of-research/agroecology-water-resilience/our-publications

Report on indigenous agroecology in New Zealand: http://www.maramatanga.ac.nz/project/indigenous-agroecology

Oxford Real Farming Conference has talks from many past conferences archived and available on its own YouTube channel, including speakers from all around the world: https://orfc.org.uk

The Biggest Little Farm is a film about the restoration of a farm in California: http://www.biggestlittlefarmmovie.com/

In Our Hands is a film made by the Landworkers' Alliance and Black Bark Films: https://inourhands.film/. It costs £6.00 to download.

Gather is a film about Native American food systems and agroecology: https://gather.film/ Agroecology voices from social movement: https://grain.org/en/article/5283-agroecology-voices-from-social-movements

Agroecology various approaches in Europe: https://www.youtube.com/watch?v=w7zqBnrLxiw

The Agricology website and YouTube channel has lots of films of farm walks in the UK: https://www.agricology.co.uk/

The Living Classroom is a film series made by the FAO about agroecology around the world: http://www.fao.org/agroecology/

Films and podcasts about agroecology in the US: www.foodtank.com

Agroecology case studies and examples from around the world can be found on the Coventry University, Centre for Agroecology and Water Resilience (CAWR) website: https://www.agroecologynow.com/vide

Films showing the work of the Magarini Organic Demonstration Farm in Kenya: www.http://www.magarini-centre.org/

1   R. Wall Kimmerer, *Braiding Sweet Grass*, Milkweed, Minneapolis, 2013.

2   M. Altieri and C. Nicholls, *Agroecology and the Search for a Truly Sustainable Agriculture*, UNEP and PNUMA, Mexico City, 2005, p. 30.

3   M. Wibbelmann, U. Schmutz, J. Wright, D. Udall, F. Rayns, M. Kneafsey, L. Trenchard, J. Bennett and M. Lennartsson, 'Mainstreaming Agroecology: Implications for Global Food and Farming Systems', Discussion Paper, Centre for Agroecology and Food Security, Coventry, 2013.

4   https://viacampesina.org/

5   Ibid.

6   J. Pretty et al., *Assessment of the Growth in Social Groups for SustainableAgriculture and Land Management*, Cambridge University Press, Cambridge, 2020.

7   A. Roy, *My Seditious Heart*, Hamish Hamilton, London, 2019.

8   https://viacampesina.org/

9   Wibbelmann et al., 'Mainstreaming Agroecology'.

10 J. Wright (ed.), *Subtle Agroecologies: Farming with the Hidden Half of Nature*, CRC Press, Boca Raton, FL, 2021. Open Access at: https://www.taylorfrancis.com/books/oa-edit/10.1201/9780429440939/subtle-agroecologies-julia-wright-nicholas-parrott

11 Wibbelmann et al., 'Mainstreaming Agroecology'.

12 Ibid.; Altieri and Nicholls, *Agroecology and the Search for a Truly Sustainable Agriculture*.

13 R. Steiner, *Agriculture*, trans. C.E. Creeger and M. Gardner, Biodynamic Farming and Gardening Association, Kimberton, PA, 1993, p. 188.

14 M. Wolfe, 'Crop Strength through Diversity', *Nature* 406, 2000, pp. 681–682.

15 Wibbelmann et al., 'Mainstreaming Agroecology'.

16 J. Pretty, C. Toulmin and S. Williams, 'Sustainable Intensification in African Agriculture', *International Journal of Agricultural Sustainability* 9(1), 2011, pp. 5–24.

17 J. Pretty et al., *Assessment of the Growth in Social Groups for SustainableAgriculture and Land Management*, Cambridge University Press, Cambridge, 2020.

18 A. Wijeratna, 'Fed Up: Now's the Time to Invest in Agroecology', International Food Security Network, Dhaka, 2012.

19 Ibid.

20 https://www.agricology.co.uk

21 http://www.magarini-centre.org/

22 https://www.fao.org/agroecology

23 https://www.landworkersalliance.org.uk

Chapter 8

Regenerative agriculture

We need to realise that mother nature is in charge. Her work can't be pre-scripted, at best we can be empathic, collaborative 'enablers' as part of an indivisible partnership for new 'nature writing' (that is, creating new farms).[1]

Charles Massy

Introduction

The term 'regenerative agriculture' was first coined by the Rodale Institute in the US and was used to 'describe a holistic approach to farming that encourages continuous innovation and improvement of environmental, social, and economic measures'.[2] The term 'regenerative' goes beyond 'sustainable' in that it suggests not only preserving resources but also improving them; the emphasis being on soil health. The term has now been used in many different contexts and organisations and refers to a form of farming and food production that is 'emergent', the term's definition not yet being fixed. The term 'regenerative agriculture' is sometimes used by industrial farmers who are using minimum or zero tillage methods, which do increase carbon sequestration considerably; these methods are still coupled to the use of nitrate fertilisers and pesticides, although in reduced amounts. The term is also used by farmers who are combining many

of the methods of farming described in this book with others to create new types of farms altogether. These groups are not exclusive, but it is a confusing picture. There is no charismatic founder, no one group that owns the term 'regenerative farming'.

If we approach regenerative farming according to the conceptualisation of sustainable food production in Chapter 2, we can understand that this is where the industrial and ecological lines of food production are converging. There are those regenerative farmers who have worked in the lineage of sustainable food systems, and there are those who have been on the industrial agriculture timeline and are in the process of changing their practice to become more regenerative while not switching over fully to sustainable systems. (By this I mean dropping the use of nitrate fertilisers and pesticides.) Chaos theory tells us that when complex systems change they go into a state of chaos, and as this transition comes to its end the new systems find equilibrium. These new systems are termed 'emergent'. We are beginning to see the new patterns that will become regenerative agriculture emerge from the chaos of change in a complex system.

The definition I have chosen to focus on in this chapter is one used by practitioners and farmers as an affiliation that has emerged from the sustainable food systems lineage. In this chapter we will explore the principles and practices that go to make up this form of regenerative agriculture and how these are integrated. It may well be that in the future there will be clear definitions, principles and practices along with standards and certification and a clear body that owns the term 'regenerative agriculture', but that is not yet the situation.

What is regenerative agriculture?

Regeneration International defines regenerative agriculture as 'farming and grazing practices that, among other benefits, reverse climate change by rebuilding the SOM and restoring degraded soil biodiversity; resulting in both carbon draw-down from the atmosphere and improving the water cycle'.[3] This regenerative agriculture also aims to be resilient to climate change and to produce higher yields of healthy or nutrient-dense food.

Charles Massy in *Call of the Reed Warbler* explains how regenerative farmers are restoring the five landscape functions: the solar energy (carbon cycle) function, the soil system, the water cycle, biodiversity, and the human and community ecosystem. To do this, farmers across the globe are using a toolkit of 'meta' tools to design and manage large and small farms to produce healthy food while restoring the five landscape functions.[4] These farmers draw on principles and practices developed over decades in the sustainable food systems described in previous chapters of this book, along with some other methods described in this chapter. What is covered here is not an exhaustive list; it's but an outline of some of the regenerative practices commonly used.

The meta tools that are outlined in this chapter include holistic farm management, keyline ploughing, conservation tillage systems and repairing the soil food web. Regenerative agriculture also uses permaculture design and agroforestry, biodynamic, organic and agroecology methods. Darren Doherty, an Australian farmer who was one of the first practitioners to use the term, refers to 300 'tools' that he uses when designing farming systems.[5]

The key thing is integration of many tools and systems, which enables farmers to regenerate the five landscape functions as described by Massy, and first and foremost the regeneration of the soil. The regeneration of farms and landscapes is a process, rather than a state; it is a journey rather than a place. The starting point of the journey upon which a farmer sets out will depend upon what kind of farmer they are already. An industrial farm can take the first steps towards repairing their soil by introducing minimum tillage systems. Organic or biodynamic farms may take their practice a step forward and weave agroforestry or mob-grazing methods into their farming practice.

Food and products can be sold and bought under the Regenerative Organic Certificate (ROC), administered by the Rodale Institute in the US.[6] The Savory Institute has created a 'Land to Market' symbol attached to an 'ecological verification outcome' audit.[7] A Greener World has set up a series of accreditation protocols in both the US and the UK for produce coming out of farms that use regenerative systems. Pasture for Life has certified pasture-fed meat in the UK.[8]

Where did regenerative agriculture start?

It is difficult to say where regenerative agriculture started, since it has grown out of the systems that have gone before it. It may have been with Allan Savory and colleagues' development of holistic grazing systems in Africa in the 1970s, or arguably with Darren Doherty, an Australian farmer who studied permaculture design and then began working and designing larger farms and weaving permaculture together with many other systems. It may have been with the Rodale Institute, which coined the phrase 'regenerative agriculture' as the next step from organic farming. It has become a buzzword that has seen rapid uptake over the last few decades.

Holistic farm management and decision-making were developed in Africa in the 1960s by Savory. Yeomans's work on keyline systems was developed in the 1940s in Australia. The soil food web was developed by Elaine Ingham in the 1990s in the US. Conservation tillage systems have been in development since the 1980s. What is common to all these is that they are often practised – in the US, Africa or Australia – on a very large scale. They are also used in Europe but on a smaller scale.

Principles of regenerative agriculture

The overarching principles of regenerative agriculture are in flux, since it is a new discipline. The principles listed by Terra Genesis International are overarching:

- improve whole agroecosystem in terms of soil, water and biodiversity;
- undertake context-specific design for farms and food systems, including holistic decisions for each farm;
- develop just and reciprocal relationships between all the stakeholders;
- continually grow individuals, farms and communities.[9]

These principles sit comfortably with Massy's framing of regeneration agriculture as referring to the regeneration of the five landscape functions: the carbon, nutrient and water cycles, restoring biodiversity and the human economic ecosystem.

Practices of regenerative agriculture

Each of the practices listed here could make a book in its own right, and indeed many books have been written about each of them. What follows is a very brief overview with signposting to more detailed texts, films and information sources.

Holistic management and decision-making

Seventy per cent of the world's agricultural area is grassland, whether that be prairie, steppe, savannah, rangeland, veldt, pampas or meadow. This is mirrored in the UK, where two-thirds of the land is grassland in the form of improved pasture, moor, commons, wetland meadow, upland grazing, flood plain, paddocks, meadow, and crop rotations. Much of the world's vast area of grassland is found in regions that have been colonised by European farmers, who have taken farming systems and livestock suited to European conditions to new soils, water systems, habitats and climates. Over the past century or so, these grasslands have been farmed using industrial farming systems. In the US the prairies were ploughed and cropped, resulting in the disastrous Dust Bowl in the 1930s. The replacement of indigenous animals with herds of sheep and cattle has also led to a process of desertification in native grassland ecosystems. The desertification was initially thought to be caused simply by overgrazing by the livestock that had replaced the native herds or indigenous animals, but was subsequently found to be driven more by the way the herds grazed than by stocking rates.

Allan Savory was an ecologist in a wildlife park in Zimbabwe that was drying out. He initially thought that this was caused by the fact there were too many elephants in the park, overgrazing it, and the park was too small to support them. He and a team of advisors advocated culling 40,000 elephants in 1969, but the process of desertification continued. This led him to discover 'mob grazing' by large herds on the plains of Africa and his mission to develop and train farmers across the world in this method. He has remorse for culling the elephants unnecessarily.

Savory noticed that the natural herds on the savannah tightly 'mob graze' a patch of grass on one day and then move on, the herd remaining closely packed for security. The tight grazing means that the grass is eaten but then has time to regrow fully and so can recover its full strength. The subsequent regrowth consists of nutrient-dense grasses. As the grass is grazed above ground, the roots die back and then they regrow as the shoots regrow. Both the dead trampled grass above ground and the roots that have died back below ground start to decay and this feeds the soil microbes and worms. The soil structure is also improved as the roots shrink and expand in the soil. The trampling herds flatten the grass to create a 'thatch'. Thus mob grazing cuts down on evapotranspiration from the soil. The grass covering keeps the temperatures more stable (warmer at night, cooler in the day), prevents damage from heavy rainfall and improves soil ecology and stability. The herd leave dung and urine behind,

providing fertility and important food for the bacteria and fungi in the soil. The working of grass roots and shoots into the soil, along with the dung, adds organic matter to the soil and this in turn increases the sequestration of carbon from the atmosphere.[10] Savory developed these observations into a land management process that is summarised in these four points:

- Nature functions in wholes and patterns. Land and resources need to be managed holistically; it is vital to remember the interconnections between land, people, livestock, biodiversity and water. All decisions need to be made with all the outcomes in mind, the ecological, the social and the economic.
- Understand the complex environment you manage. Most farmers and ranchers fight nature, but nature always wins. Sustainable solutions are much cheaper and less damaging than treating the effects and symptoms of a dysfunctional system.
- Livestock can improve land health on native grasslands and can be used as a substitute for lost keystone species. When managed properly in a way that mimics nature, agriculture can heal the land and even benefit wildlife, while at the same time benefiting people.
- Appropriate management of livestock. Control of timing is crucial. Time, rather than number of livestock, governs grassland health. Land must have time to rest.[11]

In Australia, the Americas and Africa vast farms are implementing and adapting holistic grazing practices for livestock on native grasslands. In smaller mixed farms, as are commonly found in Europe, mob grazing is used regeneratively to replenish the soils in between arable crops. Before the advent of nitrate fertilisers arable cropping land needed to be put down to 'leys', that is, grass and clover mixes in a rotation and grazed with cattle or sheep to replenish the fertility. This practice generally stopped with the advent of nitrate fertilisers. If a farm is to regenerate its soil it needs to replenish its fertility and maintain and improve the soil biome. This can be done by means of grass and herbal leys grazed by livestock. A healthy soil biome can also be maintained in a stockless system with the use of green manures and compost. Mob grazing with the cattle – moving them on in a tight pack every one to three days and giving the land a rest in between – increases the SOM and the abundance and diversity of the microflora and microfauna. This improvement of the soil biome leads to an increase in carbon sequestration. It increases biodiversity in the fields as insects feed on dung and then in turn feed the birds. Raptors turn up to eat the small mammals exposed by the short grass. The prevalence of ticks and worms in the livestock is reduced because of the rest period for the grass: by the time the stock return, the parasites have died out.

Both the Savory Institute and Holistic Management Institute offer training in these systems. In total they estimate that these methods are used on over 50 million hectares across the globe, with more than 70,000 practitioners across 130 countries.[12]

The Savory Institute has developed a marketing symbol and an accreditation process that is significantly different from the organic accreditation process. Instead of an audit trail as in the organic certification process, this process has been developed as a tool to help the farmer

learn how the farm is performing in response to their practices. The accreditation process is an impact assessment. It measures soil health by measuring the numbers of worms, the amount of bare earth, soil characteristics, the biodiversity of plants and birds, and ecosystem function and cycling on the farm. These are all part of a package of metrics with which the farmer is equipped to undertake the accreditation process to obtain the marketing symbol.

Keyline ploughing

Keyline ploughing was developed by Yeomans in Australia in the 1940s especially for arid farmland. This method can be also used in wetter regions as water resources become more erratic. A 'keyline' in a landscape is the point at which a slope flattens out, where the slope of the hill changes from convex to concave. It is at this point that water movement down a hillside slows down, water accumulates, topsoil is deposited and springs frequently occur. In European landscapes ancient farmhouses and settlements are often

Figure 8.1: Keyline plough.

Courtesy of Rural Agri-Innovation Network

found on a keyline, since this is where the water sources are found. The concept of keyline ploughing is to find the keyline on a farm or field and then use a keyline plough to plough parallel lines above and below the initial keyline.

A keyline plough is similar to a subsoiler implement, with a 'ducksfoot' at the point running through the soil. It does not turn the soil over, but rather 'rips' through the surface of the soil, 'fluffing' it up. It runs at a depth of 50cm or 18 inches beneath the surface. This eases compaction and allows air into the soil, increasing water percolation down through the soil and allowing microorganisms to thrive in the topsoil. A compost tea can also be injected deep into the soil at the same time to add in soil microbes. Water percolation through the soil is increased, as is the soil's waterholding capacity; this then helps to reduce flooding further down the valley.

In non-arid farming systems this approach may seem counter-intuitive, since most farmers are taught to put in drainage systems to remove excess water. However, allowing the water to percolate down into the deep subsoil, where it is held in organic-rich soil with plenty of air porosity, can replenish the local water cycle and watershed. With climate change and erratic rainfall patterns, many farmers are experiencing long periods of heavy rain followed by long periods of drought. Reconditioning the soil in this manner helps the soil to produce consistent crops through these extreme weather events.[13]

Yeomans pioneered this method in the arid climates of Australia. He coupled keyline ploughing with a series of dams, leaky dams and swales along the contours in the rangelands of Australia. He also planted trees along keylines and contours, completely revitalising large areas of land. We have trialled this system on our farm at Huxhams Cross in Devon (see Chapter 10).

Replenishing the soil food web.

Elaine Ingham first described her concept of the 'soil food web' in the 1990s to explain the science of soil ecology. The soil food web functions to support plant growth and animal nutrition and to combat pests and diseases, the main point being that a living healthy soil will grow healthy plants and animals.

The key element in Ingham's work was the observation that any chemical-based fertiliser or pesticide damages the soil food web. She has pioneered ways of improving the soil microbiology, composting and the use of compost teas to inoculate the soil and plants with bacteria and fungi. She has created methods of testing soils to see how intact the soil food web is in them. She has developed the use of compost teas or 'amendments' to inoculate the soil and plants, by adding compost to warm water and bubbling air through it and then spraying the resulting brew on to the plants or the soil. The rich mixture of bacteria or fungi will either eat or outcompete any pathogenic bacteria or fungi on the plants.[14]

Coupled with Ingham's work is that of Paul Stamets, who has studied fungi in the native old growth forests of Washington State – among the few remaining old growth forests in the world. What he found was a complex system of fungi growing in partnership with tree roots. Some of these he suggests are among the largest and oldest organisms in the soil. As old growth forests are clear felled these organisms are destroyed. He has found that fungi can be used to clean soil of contaminants such as DDT, which is a long-term contaminant in soils. Fungi help repair and restore damaged soils and can help trees and perennial crops in particular to grow better, by facilitating the uptake of nutrients from the soil. Recent research shows that fungal mycelia provide a 'superhighway' that moves nutrients and sugars from one tree to the next. Fungi are a key component in the decomposition process in the soil; plant debris is broken down by first fungi and then bacteria, becoming part of the organic matter that stores carbon in the soil. Mycorrhizal fungi live in the soil in association with plants and receive carbon in the form of sugars from the plants' photosynthetic process which leak out of the roots. In return the fungi collect minerals, nitrogen, potassium, phosphates and micronutrients from soil particles and feed them to the plants.[15] They also release a protein called glomalin that helps soil particles to form aggregates. These aggregates form the stable mineral and organic particles that are able to store high levels of carbon in soil for the long term.[16] Stamets (and Ingham) also suggests, that fungi can be used as pesticides and fungicides and can replace antibiotic medicines.

Beneficial fungi can 'antagonise' pathogenic fungi, reducing their ability to destroy their host plant.

Conservation tillage or no dig systems

Increasing the soil microbial activity is a key facet of regenerative agriculture. From this everything else flows. In industrial farming systems, it is the use of nitrate fertilisers and pesticides and deep cultivations, in particular the inversion of the soil in ploughing, that damages microbial activity.

Farms that stop using herbicides to kill weeds need to continue to use shallow cultivations to bury weeds even though they know that this damages microbial activity by burying it underground. Microbial activity flourishes in the top 30cm of soil where the air can easily diffuse through the soil. Many industrial farmers have improved their soils by reducing ploughing, in some cases to zero, but they still have to keep using herbicides to control weeds.

Conservation tillage has been pioneered for annual crops both in the vegetable-growing world, where it is called 'no dig', and in the arable-cropping world, where it is called 'minimum tillage' or 'zero tillage'.

Conservation tillage on a farm scale requires directly drilling the next crop through the previous crop's stubble. The farmer removes straw and grain but aims to leave 30 per cent of crop residues on the surface to protect it from heavy rain and to build up organic matter in the soil. Specially adapted machines slice through the stubble and soil, drilling the next crop in amidst the crop debris. The benefits of minimum or zero dig systems are that they increase the SOM, reduce water loss from the surface of the soil because it is covered with crop residue, increase carbon sequestration and require fewer tractor hours and so reduce fossil fuel usage. To reduce compaction by tractors, smaller tractors are often used.

Weeds can build up in this system and these are sometimes spot treated with herbicides. Nitrate fertilisers are still used but the amount required is reduced by underplanting the cereals with clovers and by the fact that the soil has more microbial activity.[18]

Massy writes about the 'no kill' methods

Zerodig at Oakbrook Farm – by Mario Peters and Christopher Upton

At Oakbrook Farm near Stroud (in Gloucestershire), one of the Biodynamic Land Trust's farms, the Zerodig approach is pioneering a new form of market gardening in the UK. As the soil biology develops through Zerodig farming the biodiversity of the land is regenerated.

Harnessing nature in this way allows Zerodig market gardens are commercially viable at small scale, sustaining rural communities.

The Zerodig beds are prepared in blocks of 10 beds and nine separating paths. Each bed is 10m long and 0.75m wide, a bed area of 7.5m². The paths between beds are 0.45m wide and each block is framed by a path 1.0m wide.

Ramial wood chips, with their high content of minerals, sugars and complex plant compounds, are made in the winter when the trees are dormant and their stored nutrients are at their maximum. Using ramial wood chips as the primary source of organic matter ensures a rich, healthy and fungal-dominant soil biology.

When the soil food web includes a strong fungal network, plant roots will connect with fungal hyphae, called mycorrhizae, which can amplify the reach of the plant roots at least 100 times, perhaps up to 1,000 times. This means the plant is more drought tolerant and can access plant nutrients from a much wider area.

Fungi also synthesise many complex biological molecules that are important for both plants and our health. For each given area of Zerodig beds, it is estimated that three to five times as much agroforestry area will be needed to provide enough wood chips to feed the soil.

Alternatively, hedgerow trimmings can be used. Food production will happen in the smaller area of the Zerodig beds, but at a very productive rate, and the surrounding agroforestry rows will also sequester carbon, provide habitat for biodiversity and act as a windbreak.[17]

Step 1: Planning the site

1. Measure and think about the site, aspect, slope.
2. Decide where to position the bed blocks on the site, leaving enough space for other activities and predator host areas.
3. Make an accurate scale drawing and mark out the beds on the ground.

Step 2: Suppress weeds and prepare paths

1. Cover marked-out beds with cardboard.
2. Cover cardboard with reversed turf from paths, cut 5–10cm deep.
3. You should have created a sandwich of soil–grass–cardboard–grass–soil on the beds and paths dug out to a depth of 5–10cm.

Step 3: Fill paths with wood chips

1. Wood chips will encourage fungal growth, help with drainage and provide a clean working environment.
2. Don't forget to have made a 1.0m wide path around the block of 10 beds.
3. Fill the paths with wood chips.

Step 4: Cover beds with compost

1. Compost will provide organic matter and – combined with the wood chips on the paths – the right medium for a healthy soil microbiology balanced for plant growth.

Step 5: Create a seedbed and top up wood chips

1. After resting for a few weeks the compost needs to be levelled to 20–30cm above the original soil and raked to a crumbly surface. Additional compost may be required.
2. Top up the wood chips on the paths to be level with the tops of the beds.
3. Each block of 10 beds uses 30m³ of wood chips and 23m³ of compost.[17]

developed in Australia by farmer Colin Seis – ones that don't need herbicides or cultivations apart from the use of disc coulters to slice the earth so the seed can be drilled in. Seis has developed machinery and methods to slice through the native grasslands in the dry time of the year in order to sow his cereal crops. These crops grow when it rains and are harvested when mature. Once they are removed, the native grassland remains underneath and is mob grazed by sheep. This particular farm restored the native grassland, from the seed banks stored in the soil, by using mob-grazing techniques. This resonates with the systems developed by Fukuoka and described in *The One Straw Revolution*.[19] To quote Seis, 'Nature will drive it for you, all you have to do is keep out of the bloody way and stop interfering and it will fix itself.'

On a small scale, as practised by Charles Dowding, no dig methods build up the soil mycorrhizae and microorganisms, suppressing weeds and creating very fertile and productive intensive systems.[20] The raised beds are 1.2m wide. They are worked entirely by hand. This means they never need to be stood on or to have tractors running over them causing compaction. They are decompacted with a broad fork and then piled high with compost and wood chips and weeded little and often. Perennial weeds are removed painstakingly by hand. Once the system is working, the beds are very productive, requiring remarkably small amounts of work.

Integrating regenerative agriculture systems in practice

When integrated into farming practice, these systems can rejuvenate dead and compacted soil after years of industrial farming. Degraded land can be first sown with deep-rooting green manure plants. The keyline plough can be used together with compost tea injectors to let air in and water out and to inoculate the soil with bacteria and fungi. A field can be mob grazed to add further manure and to improve the soil structure. Over a few years these practices will completely transform soil and its ecosystem functioning. When they are coupled with watershed renovations, tree planting or agroforestry, whole ecosystems can be repaired.

Does regenerative agriculture work?

As regenerative agriculture is both a new discipline and a complex agricultural system that has to be customised to each place, no conclusive research has been done on the entirety of this approach to date. Some research has been done on constituent parts of it.

- Minimum tillage does increase the SOM, and this in turn improves soil fertility and water-holding capacity.
- Mob grazing does increase soil carbon and the soil's water-holding capacity.
- Pasture-fed cattle provide better-quality meat and constitute a zero input system that supports biodiversity and sequesters carbon.[21]

The anecdotal evidence from practitioners is that regenerative agriculture works. You only need to watch the film *Kiss the Ground* or read Massy's *Call of the Reed Warbler* to see the stark differences between adjacent farms delineated by fence lines. There is parched grassland on one side and rich soil with verdant grassland on the other. Testimonies of countless farmers across the world who are transitioning to regenerative agriculture systems demonstrate what is possible.

Cattle versus no cattle systems

There is debate about whether cattle and grazing should be integrated into sustainable and regenerative farming systems.[23] One opinion is that, as all cattle produce methane, they should be excluded from sustainable farming systems. Moreover, in the global north an average diet includes eating up to four times more beef than is healthy, which causes obesity and heart disease. The Food Climate Research unit at Oxford University produced a report called 'Grazed and Confused'[22] which argues that it is preferable for everyone to eat less meat and to reduce livestock numbers by switching to plant-based diets. The simple answer is that people in the global north should all eat less meat and there should be fewer cattle in the world.

The second opinion is that cattle that are entirely pasture fed with a holistic planned grazing system do sequester at least as much carbon as they produce and support biodiversity in the process. Seventy per cent of the world's agricultural area is grassland, where cattle and sheep eat grass and produce meat or milk and cheese. In some parts of the world up to 95 per cent of the land will only support grass. Meat and milk from grass-fed animals are healthier than from grain-fed livestock. Research at Newcastle University shows that pasture-fed meat and milk are higher in omega-3 fats,[24] and in the film *Gather* a young Native American scientist shows how bison meat from her native prairie is better for her health than the grain-fed beef that many other young people eat, both on and off the reservations.[25]

Many grasslands are facing desertification owing to climate change. Holistic grazing systems can reverse the process of desertification, sequester carbon and restore water cycles. The exact figures for how much carbon can be sequestered per hectare vary and are contentious, but the fact that grasslands are regenerated by holistic grazing practices is not disputed. In the mixed farming systems of Europe, the Sustainable Food Trust suggests, mob grazing combined with arable cropping systems do sequester at least as much carbon as they produce and do replenish the soil biome. If herbal leys are used in rotations with arable crops, then grazing the herbal leys with livestock not only produces more food (milk and meat) but also contributes to the farmers' income.[26]

The complex answer involves the quality and quantity of the cattle and the resulting beef coming out of the industrial food system. To put this in context, calves born from the dairy herds or specifically for beef production are weaned, and then placed in feedlots for finishing. These feedlots are where cattle are held together in a 'lot' with no grass or grazing and little space. These are found across the world but particularly in the grain belt of the USA, in Brazil and China. Feed for these cattle is grain, waste products from oil production such as soya oil cake, and fodder that is shipped to the feedlots. These will have been grown using nitrate fertilisers and pesticides, the cattle are sometimes given prophylactic antibiotics to increase their weight. The cattle in turn emit methane and produce a lot of concentrated manure. The resulting beef is arguably not as healthy for human consumption as grass-fed beef. This beef is then shipped around the world for consumption and is often very cheap.[27] People in many cultures eat more of it than is healthy for them. Because of this, many people in the world are switching to a plant-based diet. However, vegan food is often highly processed, using products such as soya and palm oil that are grown in cleared rainforest. Locally produced grass-fed beef is often a culturally appropriate and healthy regional food.

Eating it should not be confused with eating beef that has come from the industrial food system.

What do regenerative agriculture systems look like?

Both the Savory Institute website and the Holistic Management Institute offer a series of short films featuring case studies all over the world where regenerative agriculture has been introduced very successfully. In the UK the Agricology website has many examples of regenerative agriculture.[28] Massy describes the complexity of these management systems and elaborates how they work in fascinating detailed case studies.[29] Huxhams Cross Farm integrates many methods of regenerative farming (see Chapter 10).

Case study:
Gothelney Farm, Fred Price

Gothelney is a 250-acre family farm just outside Bridgwater, at the foot of the Quantock Hills. I also rent Clayhill Farm, 130 acres, on a six-year business tenancy. I took on the farm in 2009, initially embracing the mainstream narrative of an industrial and commoditised agriculture. My farming journey has brought me to agroecology and regenerative agriculture and

has been transformative both practically and psychologically, a process of self-discovery and self-preservation. Why are we told: turn left for food production and right for ecosystem health? It is in the productive grey areas in between these binaries from which a more beautiful, complex and diverse human-scale food system based on trust, relationships and community emerge. Gothelney is an altogether more woolly and happier farm for it.

The farm grows grains, raises pigs, and is now home to Field Bakery, run by Rosy Benson, that sells bread and flour to the local community and runs training courses in baking with local grains.

Connecting and feeding our community is our ultimate goal, although we also supply restaurants in London and bakeries and mills in the southwest of the UK. Sidestepping commodity agriculture has been critical in providing me with the agency to decide how we farm. Commodity systems cannot incorporate the values that add up to appetite. Rather they measure things which are convenient (price and quantity) and ignore those that are not – externalising many of the costs associated with an intensive food system. Growing a commodity fundamentally focuses the farmer on yield, at the expense of other considerations. None of which constructs a healthy food system which is good for farmers, ecology or community.

Our farming system is very much informed, inspired and borrowed from nature (and other farmers who think the same way). Earning the right to farm without inputs means considering the soil as a vibrant, living ecosystem. Our rotation and the way we grow crops attempts to reflect this. Eight per cent of the farmed area is down to habitats, refuge areas for those beneficials who will reach far into our arable fields and buffer the natural occurrence of pest and disease. But agroecology and regenerative agriculture go beyond making space for nature. It is a kind of farming which is resilient because of its deeper connection to it. I discovered that connection because of soil, and the way the farm looks is an attempt to recognise soil as a living, breathing thing, and has led to a focus on diversity and carbon.

Growing and grazing diverse leys over two years is an attempt to build our soil carbon – the foodstuff of bacteria, protozoa and microbes, the base of the food web. These are the very organisms whose lifestyles, death and decay create a habitat and cycle nutrients, the conditions for growth upon which a more extractive part of our system depends. Over the past five years we have seen soil organic matters rise by up to five per cent under this approach of no synthetic, perennials and green covers.

How do we pay for these positive parts of our system? For me this is where livestock come in. We raise pasture-based pigs that graze from April to September and run over the green cover between arable crops over the winter. Other than forage, the pig diet is topped up by cereals and pulses also grown on the farm. The absence of soya and bought-in minerals in the diet led to quite a substantial increase in finishing time – from eight to twelve months. We have begun to breed a Tamworth X Saddleback dam and put her to a Duroc sire for our finished pigs, the hybrid vigour has reclaimed some of these productivity losses.

Annual crops are, for me, the extractive moment in our system. Therefore, the way we do that 'extractive' part of our rotation matters a great deal. I've worked hard to blur the boundaries between the two sides of our rotation, moving away from monoculture cereals by undersowing and intercropping. Reducing our dependence on the plough and learning to read, love and understand weeds have been largely mind-set changes that have been hugely rewarding in practice.

Part of this refocusing has fostered a fixation on cereals that are suited to organic, agroecological systems. That means finding varieties that are taller and that have a habit of coping for themselves and growing them as mixtures or (better yet) populations so that they may express and adapt both within and across seasons to the vagaries of weather and climate.

Typically, there is a direct correlation between varieties with a high harvest index (the ratio of grain to total plant dry matter) and dependency on synthetic interventions (fertiliser, herbicide, fungicide, you name it). But as I work through around 8000 different cereal lines sourced from gene banks across the world (a project with Colin Gordon of Inchdown Farm), I'm more and more encouraged by lines with a good balance of both yield, quality and ability to cope without synthetics.

How we sell the grain matters too. Imagining an alternative grain economy with millers and bakers in the southwest of the UK (SW Grain Network) and beyond (UK Grain Lab), has been the biggest change, the most difficult, and the most transformative, and the most rewarding. It has been to move away from a commodity, big food system towards something that is human scale, about nourishing people, and built on relationships and collaboration rather than competition and protection. I believe such a system can, and in many cases does feed the world. But not by getting bigger, rather by talking and sharing – increasing the scope rather than the scale of what we do.[30]

Case study: Lynbreck Croft

Lynbreck Croft is a 60-hectare (150 acre) croft in the Cairngorm National Park in Scotland. Crofting is a part of the traditional way of life in the Scottish Highlands and one that Lynn Cassells and Sandra Baer are recreating in a regenerative model.

They bought the croft in 2016 with no previous farming experience, although they were both working in conservation and rewilding in previous jobs. Their aim at Lynbreck is to enhance the biodiversity on the farm whilst producing food and a living for themselves. They refer to themselves as 'soil builders, tree planters and growers of food and minds.' In their own words they say that:

Our goal is to understand and work with natural processes to produce the best food we can. If it doesn't work for nature, it doesn't work for us. Our goal is always to build biodiversity and have healthy soils. The long-term health of our land is our greatest asset. The animals and insects that live here are a part of our team. It is our duty as guardians of the land to ensure they lead the best lives that they can – to express their natural behaviours, to live a low stress life and to have the food, water and shelter that they need. Our goal is to produce food of exceptional quality to nourish our local community. We acknowledge the importance of direct sales and sustaining local food supply chains as this provides the greatest resilience to our business model. We play a role in a global society as story tellers and educators, but prioritise our food for those around us. Our mission is to help reconnect people with the land, the food they eat, the water they drink and the air they breathe. We believe that solution-focused messages drive the long-term changes that we need to heal ourselves and our relationship with our planet.

Lynn and Sandra raise pigs, cows and hens on the farm, as well as bees. The breeds chosen can cope with the extreme weather in the North of Scotland. The hens live either in a static or in a mobile hen house (the 'Eggmobile'), spending long days scratching and dust bathing. Oxford Sandy and Black weaner piglets are bought in at different times of the year – the breed is hardy, hairy and good-tempered. They break up dense clumps of ground vegetation, creating bare patches for tree and pasture seedlings to set and grow.

The fold of Highland Cattle is moved around the farm using holistic-planned grazing which helps with the regeneration of grasslands and woodlands. They are 100 per cent pasture and tree leaf fed, incredibly docile, and they graze, trample and dung the fields, their impact helping to build organic matter into the soil.

The paddocks are given a long rest period to allow vegetation to fully recover and regrow. In the winter they are fed pasture hay as well as bundles of tree and nettle hay to supplement the forage in the fields. Bale grazing targets specific areas in the winter that need greater diversity.

In total they have planted nearly 30,000 trees creating new native woodland, hedgerows, shelterbelts and a future crop of trees to make into tree hay for animal fodder. Nine hectares are set aside for natural regeneration of woodland more trees are being set into fields including fruit and nut varieties.

All the produce is sold directly to the customer with meat and eggs exclusively to a local market. Lynn and Sandra also offer tours of the farm and run a number of on farm courses. Their book on 21st-century crofting, *Our Wild Farming Life*, is due to be published in spring 2022.

Where is the regenerative agriculture movement going?

The regenerative agriculture movement is new and is still working out how to define what it is and what it is not. Different groups are forming networks and collaborations, with different emphases, whether it be minimum tillage, holistic grazing, large scale or small scale.

However, the organisations involved in regenerative agriculture have clear goals of consolidation and expansion. It has high take-up by farmers.

Training in regenerative agriculture

University degrees at undergraduate and postgraduate level in regenerative agriculture are now probably the most available form of state training in any kind of sustainable agriculture and are offered around the world. In the UK regenerative agriculture is offered at Writtle College, part of Essex University; the Royal Agricultural University, Cirencester; and Schumacher College in Devon. In the US the Universities of California and Oregon offer training in regenerative agriculture. The Apricot

Centre at Huxhams Cross Farm offers a two-year apprenticeship, state funded, in regenerative land systems at levels three and four.

Training or CPD for existing farmers exists in many forms. Doherty's website Regarians offers webinars, and training manuals that cost A\$20 each.[31] The Savory Institute offers seminars and webinars, and there are 47 regional hubs around the world where courses can be taken in person. The introductory course is nine days long. In the UK this is offered by Christopher Cooke of 3LM.[32]

The Agricology website offers farm talks and seminars and webinars in the UK, and there are many similar organisations around the world.[33] IFOAM runs a five-day course for farmers.[34] Groundswell is the annual conference for farmers who practise regenerative farming in the UK.[35]

Stamets' approach to fungi is found on his website, where he offers online seminars and webinars on the use of fungi in farming systems and more.[36] Ingham's website offers webinars and courses around the world.[37]

The Regenerative International website offers access to hundreds of webinars from around the world which reflect the 'toolkit' approach to this type of farming. It lists and promotes courses and webinars on topics from beekeeping to grassland management. They have speakers such as the well-known North American rancher Gabe Brown and links to the US biodynamic conference. The website reflects the breadth of the regenerative agriculture movement.[38]

Short courses, apprenticeships and internships are run on many farms around the world, such as Richard Perkins's farm Ridgedale in Sweden and Polyface Farm in the US.[39] ∎

In a nutshell

- Regenerative agriculture is all about regenerating the soil, water cycles, and biodiversity, sequestering carbon, producing healthy food and creating economically viable livelihoods.

- It is practised on a farm scale and often on very large farms across the world.

- It uses a number of meta tools to create unique holistic management plans for farms.

- Regenerative agriculture farms recondition the soil by regenerating the soil microbiome, using mob grazing, agroforestry systems, minimum tillage and keyline ploughing. Permaculture design can be used to weave these tools together in a complete farm design.

- 'Regenerative agriculture' is a term that is accessible to many farmers and it is also a term that academics and policymakers can access and use.

Further reading, links and films

R. Perkins, *Regenerative Agriculture*, self-published, 2019.

G. Powell, 'Sustainable Grazing Strategies Meet Ecological Demands', Nuffield Scholarship Report, 2018.

G. Brown, *Dirt to Soil*, Chelsea Green, Hartford, VT, 2018.

D. Montgomery, *Dirt*, University of California Press, Berkeley, 2012.

D. Montgomery, *Growing a Revolution*, W.W. Norton, New York, 2018.

J. Stika, *A Soil Owner's Manual*, self-published, 2016.

V. Shiva, *Soil Not Oil*, Zed Books, London, 2016.

Keyline Designs: www.keyline.com.au

Allan Savory TED Talk: https://www.ted.com/talks/allan_savory_how_to_fight_desertification_and_reverse_climate_change?

How to make a Zerodig bed: www.zerodig.earth

Probably one of the best-known films about regenerative agriculture is *Kiss the Ground*: https://kisstheground.com/

The incredible Vandana Shiva appears in *The Seeds of Vandana Shiva*: https://vandanashivamovie.com/

How Is Regenerative Organic Farming Better is a film by John Kempf in the US.

From the Ground Up: Regenerative Agriculture is a film about the story of regenerative agriculture in Australia.

Keyline Design Explained on a Beach is a film by Darren Doherty on YouTube.

Oregon State University Ecampus has online training in keyline management. There is a whole series where you can see Yeomans farms and Yeomans ploughs in action.

Fantastic Fungi is a film by Paul Stamets: https://fantasticfungi.com

Films about no-dig systems may be found on www.Charlesdowding.org

Symphony of the Soil is a film about the soil food web: www.symphonyofthesoil.com

1 C. Massy, *Call of the Reed Warbler*, Chelsea Green, Hartford, VT, 2017.

2 https://rodaleinstitute.org/

3 https://regenerationinternational.org/

4 Massy, *Call of the Reed Warbler*.

5 http://www.regrarians.org/

6 https://rodaleinstitute.org/

7 https://www.savory.global

8 https://www.pastureforlife.org

9 https://www.terra-genesis.com/

10 https://www.savory.global

11 Ibid.

12 Ibid.; https://holisticmanagement.org

13 P.A. Yeomans, *Water for Every Farm*: *Yeomans Keyline Plan*, Griffin Press, Adelaide, 1993.

14 https://www.soilfoodweb.org

15 P. Stamets, *Mycelium Running: How Fungi Can Help Save the World*, Ten Speed, Berkeley, 2005.

16 J. Moyer, A. Smith, Y. Rui and J. Hayden, 'Regenerative Agriculture and the Soil Carbon', Rodale Soil Carbon White Paper, 2020.

17 https://www.zerodig.earth

18 *Global Overview of Conservation Agriculture: Principles, worldwide spread and main benefits*: https://groundswellag.com/wp-content/uploads/2019/10/Groundswell-2019-PPT-24-06-19-Final.pdf

19 M. Fukuoka, *The One Straw Revolution*, New York Classics, New York, 2009, published 1978.

20 Massy, *Call of the Reed Warbler*.

21 C. Dowding, *Organic Gardening the Natural No Dig Way*, Green Books, Dartington, 2013.

22 https://www.pastureforlife.org

23 T. Garnet et al., 'Grazed and Confused: How Much Can Grazing Livestock Help to Mitigate Climate Change?', FCRN report, 2017.

24 D. Srednicka-Tober et al., 'Composition Differences between Organic and Conventional Meat: A Systematic Literature Review and Meta-Analysis', *British Journal of Nutrition* 115(6), 2016, pp. 994–1011.

25 https://gather.film/

26 https://sustainablefoodtrust.org/articles/grazed-and-confused-an-initial-response-from-the-sustainable-food-trust

27 Paul L. Greenwood, 'Review: An Overview of Beef Production from Pasture and Feedlot Globally, as Demand for Beef and the Need for Sustainable Practices Increase', 2021, 100295, ISSN 1751–7311, https://doi.org/10.1016/j.animal.2021.100295

28 https://www.agricology.co.uk

29 Massy, *Call of the Reed Warbler*.

30 https://www.gothelneyfarmer.co.uk/; https://southwestgrainnetwork.co.uk/; https://www.ukgrainlab.com/; https://www.fieldbakery.com/

31 http://www.regrarians.org/

32 https://www.savory.global

33 https://www.agricology.co.uk

34 https://www.ifoam.bio/regenerative-agriculture-building-foundations/networks-and

35 https://groundswellag.com

36 https://fungi.com/

37 https://www.soilfoodweb.org

38 https://regenerationinternational.org/

39 https://www.Ridgedalepermaculture.com; https://www.polyfacefarms.com

Part 3

bringing **about the**
sustainable
agricultural
revolution

Chapter 9

Meeting the challenges to sustainable food production

We cannot solve our problems with the same level of thinking that created them… The only time we ever truly transcend a problem is when we get new, fresh thinking about it. In other words, we experience a jump in consciousness.[1]

Albert Einstein

Introduction

In Part 1 the four main challenges facing food production in the 21st century and the thinking that got us there were explained. Part 2 has described the toolkit of sustainable food production systems that are available to farmers and growers and can be woven together to meet these challenges. The sustainable food systems arose in response to the industrial food system, and these two systems have followed divergent paths since 1909 and the development of nitrate fertilisers. Part 3 describes how the toolkit of sustainable food systems can be used as part of a systems rethink of farming and food production systems.

This means going back to the point of divergence and recreating truly modern farming systems using the toolkit of sustainable methods that are now available to us. If we return to the metaphor of the 'fringe arts festival', we can see that this festival has been going on for almost 100 years; the expertise and knowledge, the skills and the science have been building up so that we can now use these systems and know that they work. To quote David Holmgren, we can now 'design the world we want rather than fighting the world we don't want'.[2]

Chapter 9 explores how the toolkit of sustainable food systems can meet the food challenges. Chapter 10 illustrates how we have integrated these systems and implemented them at Huxhams Cross Farm, offering some early results of the environmental, economic and social impact of the first five years. Chapter 11 explores how a transition from industrial to sustainable food systems might come about.

All of the food systems outlined in this book have similarities, but they contribute different qualities to the toolkit of sustainable food systems. To put it another way, together they offer a holistic approach to sustainable food production.

To reiterate what was introduced in Part 1, the four main challenges facing food production in the 21st century are climate change mitigation and adaptation, biodiversity loss and producing enough healthy food for an increasing population from the same amount of land. The corollary is that we need to create farms that produce lots of high-quality food for local people while supporting biodiversity, that both are low on carbon usage and sequester carbon and that can cope with challenging weather events. At the same time, the farms need to be financially viable for the farmers, to sell food at affordable prices and to allow access to people for their health and well-being.

This chapter is not a piece of academic research, but observations and experiences from a practitioner, backed up by some research findings. There is plenty of evidence to support the contrary position, that industrial farming can produce the most food and solve lots of the challenges facing farming in the 21st century. I would encourage everyone to read around the subject, visit farms, watch relevant films, eat the food and make their own decisions. It is my opinion, however, that the toolkit described in this book has the ability to meet these challenges fully and directly over the next 30 years.

Sustainable food systems in the 21st century

The kinds of farms I'm talking about – farms for the future – reduce their direct carbon emissions by reducing their consumption of fossil fuels in tractor use, cold stores, long distance or airfreight deliveries, and heated glasshouses. They reduce some of their embedded carbon and GHG emissions by eliminating the use of nitrogen-based artificial fertilisers and pesticides. This also allows the soil biome to recover and regenerate.

The food production systems instead rely on localised means of maintaining soil fertility such as manure, green manure and good soil management. Pesticides are replaced by functional biodiversity, genetic resistance to disease, crop timing and crop diversity, and the use of fleeces or other crop covers that provide a barrier to pests yet still allow the crops to grow. Herbicides are replaced by mechanical weeding or hand weeding. Fossil fuels are still used to fuel tractors and machinery and in multiple-use plastic equipment such as crop covers, which can be reused for many years.

In farms for the future, food supply chains are relocalised and seasonal to reduce food waste. Plastic packaging is reduced, or replaced with cellulose packaging or plastic bags that can be used multiple times. Meat production is shifted towards poultry and away from beef.

Any cattle and sheep that are reared are pasture fed, either on permanent pasture or on mixed farms to regenerate soils after arable crops. More plant-based protein is produced through growing legumes such as lentils, beans and pulses that can be grown, dried and eaten locally.

Farms for the future increase their ability to sequester carbon in their soil in the form of organic matter, as well as planting more trees to soak up carbon and lock it into their wood. The trees also reduce the soil erosion and flooding caused by extreme weather conditions.

Greater amounts of organic matter in the soil and larger numbers of trees will enable farms to adapt to climate change as the weather becomes more erratic and unpredictable. The farms and food systems are resilient and diverse; their land can soak up excess water and cope in dry conditions, as well as providing food no matter what the weather. The soil is also more resilient to weather extremes when a greater diversity of crops are grown. Farms for the future are polycultural, rather than monocultural. This kind of farming imparts resilience to both food security and farm income.

Farms for the future will repair the soil biome, replenishing its organic matter, and so will support biodiversity from the soil upwards. The farmland habitats of hedgerows and field margins join up to small woodlands, ponds and damp meadows. Wildlife requires habitat in both summer and winter. Many species either hibernate or overwinter in the form of eggs or pupae; either way, they need to be tucked away in a warm place hidden from predators. All creatures need nourishment, and the food chain for biodiversity on a farm starts with the soil biome. The bacteria and fungi break down plant matter to provide food for the worms, the birds and the trees. When farmers repair the soil ecology, everything else follows. Clovers and flowering trees are planted to improve soil fertility as well as to feed bee populations.

Conservation grazing in wild meadows allows plant populations to recover. Crops such as brassicas are allowed to flower in the fields to feed the bees.

Farmers harness 'functional biodiversity' on their farms, that is, predatory insects and birds to control the pests on the crops, in place of pesticides, and other insects to provide pollination. The sustainable farmer values and enjoys biodiversity and can see the worth of having it on the farm. It serves the farmer well to provide habitat and food for many species.

Farms for the future are highly productive, and collaborative. They range from micro-scale urban farms and gardens, to medium-sized peri-urban farms, to larger-scale farms outside in rural areas. Some food is shipped over land or sea in different seasons, but never flown.

On the farms of the future, food waste will be reduced by utilising any gluts of produce and any graded-out produce that cannot be sold because it's too small, too big or just plain wonky. This applies with grain, milk, meat, fruit and vegetables. With the right expertise processing rooms and on the right scale this unwanted food can be processed and gain added value through being made into jams, chutneys, juices, cheese, pies. Small grains are used as animal food, as are fruit and vegetables that have gone off and can't be processed. The health dividend of people eating all this good food will be huge. People will use their local farms and gardens as well-being and health centres. For these changes to happen on a larger scale, access to land will need to be more equitable, and working as a food producer will be a well-paid and valued career.

How does the sustainable food systems toolkit meet the four challenges?

How do the food systems described in this book fare in relation to the four challenges? What is the scientific research and evidence? There is far more research on the sustainable food systems than I could review in this book. Those pieces of research I have chosen are themselves reviews of research, and ones that clarify and provide insight into the functionality of how the respective systems actually work. I have attempted to simplify the research to make it accessible.

1. Mitigate climate change

Sustainable food systems have lower carbon emissions both directly and in embedded forms. They also sequester more carbon in their soils than do their industial counterparts. They have lower carbon emissions because they do not use nitrogen-based fertilisers or pesticides.

Emissions from nitrogen-based fertilisers on their own account for some six per cent of worldwide carbon emissions.[3] Sustainable farming systems, on the other hand, have pioneered the development of green manures, functional biodiversity, crop timing, crop diversity and mechanical weeding methods to produce local, seasonal, delicious food without the use of fertilisers or pesticides. These are multifunctional systems. The use of clover-rich green manures replenishes nitrates in the soil, increases the SOM, increases the soil's biodiversity, provides bee fodder and provides grazing for livestock – and in doing so produces food. Agroforestry rows improve water storage and carbon sequestration, provide habitat and produce food.

Most of the sustainable farming systems aim to be closed loop. The biodynamic, permaculture and agroecology systems in particular have this as one of their core principles. Holistic grazing systems have pioneered purely grass-fed cattle-grazing systems, which reduce the need to import industrially grown feed for the cattle and to export the meat produced.[4]

Sustainable farming methods have lower carbon footprints because they use 30–50 per cent less energy per tonne of yield than industrial farms when the food system is examined in its entirety, as shown in long-term trials by FiBL.[5] A common criticism of sustainable farming methods is that they use more tractor hours, and therefore more fossil fuel, to carry out the mechanical work required to control weeds, but when this is examined against the footprint of using herbicides to control the weeds the organic and biodynamic farms perform the best.

Sustainable farms generate renewable energy. Permaculture-designed farms are multifunctional; they are designed to be micro power plants generating renewable energy from solar panels on barn roofs, wind turbines, aerobic digestion plants and, possibly in the future, combined heat and power (CHP) systems using biomass. These power sources go towards replacing the need for electricity produced from fossil fuels to power cold stores, delivery vehicles and heated offices. Permaculture design has developed glasshouses that are heated by storing heat accumulated in the day in water tanks and the soil, as well as biomass boilers that are fueled with wood pellets sourced from agroforestry rows. These glasshouses can grow out-of-season and temperate crops, obviating the need for airfreighted out-of-season crops or out-of-season crops grown in glasshouses heated with fossil fuels.[6]

The use of fossil fuels is reduced by creating shorter supply chains. Localised food systems have been pioneered by biodynamic farms and are commonly used by organic farms, under the heading of 'community-supported agriculture' (CSA). Linking the produce from the farm directly to the consumer reduces food waste, food miles, airfreight, refrigeration and the use of plastic packaging to virtually zero. It encourages the consumer to eat a local seasonal diet, thereby reducing the need for imported food.[7] There is less food waste because the customers are happy to take slightly misshapen produce and, as sustainable farms are often mixed farms, the grade-outs are often fed to livestock or composted, creating a circular food economy.

At a political level the concept of food sovereignty, the right of people to maintain their lands and their own food culture, to grow and eat locally produced healthy food in the face of the dominance of agribusiness, has been key in helping millions of farmers and smallholders across the globe to have a political voice. They have been enabled to retain their local food culture and systems, and either not to transition to industrial systems in the first place or to transition from an industrial system to an agroecology model while increasing their yields. The agroecology movement is the largest of all the sustainable food systems and its social and political methods have made it the most effective means of grassroots change to cut carbon emissions.[8]

When it comes to mitigation of climate change, organic farms sequester more carbon than do industrially farmed soils. The Soil Association's review of evidence estimated that organic farms sequester two tonnes of carbon dioxide per hectare in their soil per year. This is 28 per cent more than with an industrially farmed soil. Biodynamic farms sequester 25 per cent more carbon than do organic farms.[9] Agroforestry systems increase carbon sequestration yet further; it is estimated that in Northern Europe an agroforestry system

can sequester as much as four tonnes of carbon dioxide per hectare per year and up to six tonnes per hectare per year in tropical climates.

One possible reason for the biodynamic and agroforestry soils being so good at sequestering carbon is that they have a particularly high level of fungi. Soils with a high fungi to bacteria ratio perform well in carbon sequestration and water-holding capacity because of the mycorrhizae that live symbiotically with plants, absorbing carbon exudates from their roots and supplying them with up to 80 per cent of their nitrate and phosphate requirements. Mycorrhizae secrete glomulin, which helps form soil aggregates, and these aggregates sequester carbon for the very long term. Mycorrhizae thrive in soils rich with perennials, deep-rooting plants and a wide diversity of plants. They can be added to the soil in the form of amendments (as prescribed by Elaine Ingham's soil food web practice), compost or biodynamic preparations.[10]

Calculations at Yatesbury Biodynamic Farm in Wiltshire in the UK, using the farm carbon toolkit, have found that the farm sequesters nine times as much carbon as it uses, 19 tonnes per hectare.[11] At my own farm, Huxhams Cross, again using the farm carbon toolkit, we found that we sequester twice as much carbon as we use; almost 10 tonnes of carbon per hectare per year are sequestered (see Chapter 10).

The Soil Association review (written in 2009) estimates that if all the farms in the world were organic the organic soils could sequester 11 per cent of the global total of anthropogenic carbon emissions.[12] The more recent Rodale Institute 'soil-carbon' white paper extrapolates early results from

Mechanism for carbon sequestration into the soil – by Rachel Bohlen

Building up soil carbon can help cut GHG concentrations in the atmosphere, as well as improving soil quality in a number of ways; it gives soil structure and stores water and nutrients. Carbon in the soil combines with oxygen, hydrogen, nitrogen and other compounds, forming soil organic matter (SOM). This material is amazingly complex, made of thousands of different chemical compounds that remain from the decomposition and transformation of plants, animals and microorganisms.

Adding to the complexity, carbon can be found in different physical states within soil.

It can be dissolved in water, present as larger chunks or 'particulates', enveloped by soil particles or bonded to minerals. These various forms all behave differently, and ultimately have very different impacts on plant growth, soil structure and carbon sequestration.

The soil science community has been studying this question for decades. One key distinction can provide an underlying framework for soil carbon management: particulate organic matter (POM) versus mineral-associated organic matter (MAOM) (Figure 9.1).

POM and MAOM are created by different processes and respond differently to management practices. In addition they have different potentials to sequester carbon. MAOM saturates, which depends on the amount and type of mineral in the soil, whereas POM does not. Most soils are below saturation so there is a high potential for sequestration.

POM contains partially decomposed organic fragments, for example, tiny bits of leaves or roots. POM is freely available to microorganisms, so it gets broken down much faster than MAOM. It's also vulnerable to agricultural practices such as tillage that disturb the soil. Although it contains less nitrogen per unit of carbon, the nitrogen is more readily available. POM is more readily available than MAOM, but its usefulness or quality for decomposers varies.

MAOM consists mostly of microscopic coatings on soil particles, derived from bodies and byproducts of microorganisms and plant compounds. Because it is stuck to soil particles, it tends to stay there for a long time. MAOM contains more nitrogen per unit of carbon than POM. It is sometimes called 'soil aggregates'.

Managing SOM stocks to effectively address global change challenges requires deep understanding of SOM formation, persistence and function. Since POM and MAOM form, persist and function in different ways, conceptually separating them is key to crafting effective global climate change mitigation strategies involving SOM sequestration and functioning.[15]

carbon sequestration on regenerative organic agriculture farms. It suggests that if the entire world's farming systems on both cropland and pastures were moved to a regenerative model then they could draw down more than 100 per cent of the world's annual carbon emissions.[13]

Holistic management systems suggest that planned grazing systems could increase carbon sequestration and prevent desertification around the world.[14] There is no doubt that sustainable food systems sequester more carbon, but exactly how much is still uncertain.

2. Climate change adaptation

Sustainable farming systems are more resilient to climate change than are industrial farms, owing to the increased SOM they create and the amount of surface cover they have in the form of mulch, or thatch on grassland. The organic matter increases the water storage capacity of the soil, enabling it to soak up water in times of heavy rainfall and hold on to it during periods of drought, especially if it is rich in fungal activity.[16] Keyline ploughing facilitates water penetration deep into the soil. Mulched soils in cropping areas, minimum tillage arable systems and mob-grazed grasslands prevent soil erosion and compaction during heavy rainfall.

Agroforestry trees, especially if planted on the contour, improve water penetration deep into the soil and help with flood alleviation. The trees also provide shade and wind protection to crops, reducing evapotranspiration from the soil surface. Trees can be planted for coppicing that can be used for biomass fuel or energy generation or to further increase the SOM in the form of ramial wood chips (see Chapter 8).[17]

Permaculture and agroecology farms in principle build rainwater-harvesting systems and store the water in ponds in the landscape. Biodynamic farmers have pioneered grey and

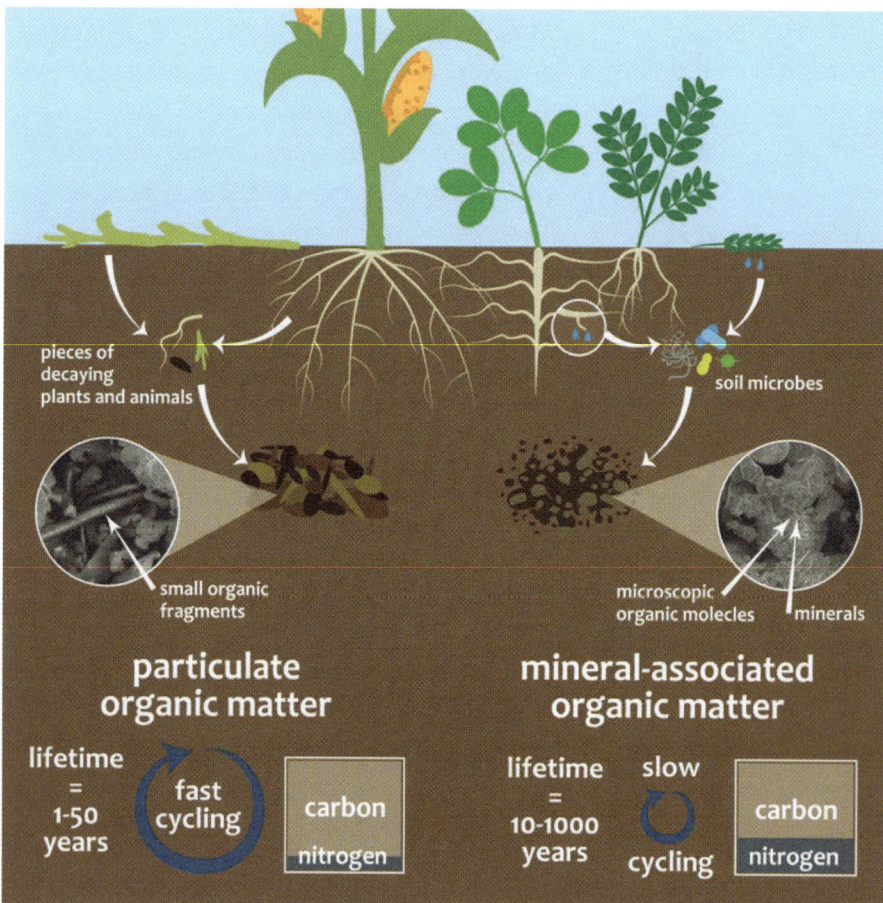

pieces of decaying plants and animals

soil microbes

small organic fragments

microscopic organic moleces   minerals

particulate organic matter

mineral-associated organic matter

lifetime = 1-50 years   fast cycling   carbon   nitrogen

lifetime = 10-1000 years   slow cycling   carbon   nitrogen

Figure 9.1: Particulate organic matter (POM) versus mineral-associated organic matter (MAOM).

black water systems that recycle water on a farm so it can be used for irrigation.

Food production can be made resilient to climate change and erratic weather by increasing the diversity of cropping. If a farm is growing 100 varieties of fruit or vegetables sequentially then there is more chance that some crops will withstand exceptional weather conditions. The use of open-pollinated seed and population crops augments the resilience of a crop to different weather events. This type of cropping plan is exactly what farms adopt if they are supplying local customers with a range of seasonal local food all year round via a CSA or a farmers' market. This concept of increasing diversity to increase resilience in the farm system applies to livestock as well, the farms rearing more types of animals and more resilient breeds suited to the local environment and selling produce via a CSA or local market. CSA and direct marketing methods have been pioneered by biodynamic farms and have been taken up by many types of sustainable farming.[18]

Sustainable farming systems relocalise the small (or local) water cycle. This practice, sometimes called 'rehydrating a landscape', is a key part of the regenerative agriculture movement. Many industrial farmers remove water from a landscape by putting in drains and removing trees. As the organic matter in the soil decreases, the soil loses its water-holding capacity and dries out. The small water cycle can be restored to a local region, or even an individual farm, by slowing down the movement of water away from the farm or farms by increasing the SOM, which behaves like a sponge. The planting of trees, including in agroforestry rows, to provide shade and reduce wind speed, together with mulching and keyline ploughing enable water to remain in the soil. The water is held in the soil and evaporates from the plants, thereby cooling the air. This water then condenses and falls as rain locally.[19]

3. Offsetting biodiversity loss

All of the sustainable food production systems support higher levels of biodiversity.[20] They are 'porous' to biodiversity. In order to thrive, biodiversity – that is, all living things – needs both food and habitat. If habitats and food sources are provided on farms then wild organisms will move in, farms ecosystems can be created and, in effect, the edges of farms can be rewilded. It usually takes five to seven years for the full farm ecology to re-establish itself as a farm transitions out of an industrial farming system. When this has happened, it brings many benefits to the farmer as well.

A review of evidence by the Organic Research Centre in the UK states that organic farms support 20 per cent higher levels of biodiversity than do industrial farms. That means greater diversity and abundance of plants, insects, spiders and worms. The soil contains more microbial diversity and abundance and higher levels of mycorrhizae. There are more bird species, especially ones that eat insects; more small mammals like mice, voles and bats; and more trees and hedgerows, owing to smaller field sizes.

Sustainable farms provide more food for biodiversity first and foremost by repopulating their soil with a fully functioning soil biome. The soil biome is restored by adding organic matter, which is the food source for soil microflora and microfauna, and by eliminating the use of fertilisers and pesticides, which kill the soil microflora and microfauna. All of the sustainable systems have higher levels of SOM and soil biological activity. The soil biome is the foundation of the food chains that build up through the ecosystem. Biodynamic farms in particular have been proven to have the highest levels of soil biodiversity among all the sustainable food production systems, with a consequently higher number of worms and beetles.[21]

Many of the sustainable farms are in reality small mixed farms that include grazing livestock on grass leys. Grazing animals will provide food for insects and worms in the form of manure. The insects and worms are then food for birds. Green manures themselves are full of clovers and buckwheat, which will be rich with bee food if allowed to flower. Careful management of vegetable crops can offer rich sources of food for bees, as when brassica crops are allowed to go to flower when most of the crops have been harvested.

Secondly, farm habitats are improved on sustainable farms. Planting agroforestry rows across a farm field provides both habitat and food for birds, worms and insects and links up fragmented habitats such as hedgerows and small woodlands for small mammals and insects.[22] The way in which these habitats are designed and managed can increase biodiversity. Agroforestry trees can be chosen so that their flowers support bee populations, and the understorey can be planted to support different insects or birds. Agroforestry trees can be also chosen for their livestock-deworming qualities. The livestock then have no need to be treated

with chemical dewormers that subsequently kill the worms in the soil.

Holistic grazing improves the habitat for wild plant populations in grasslands; the native plants thrive under this grazing regime. In pastures that have previously been intensively grazed, the fast-growing introduced grasses thrive and outcompete the native wild plants. The tight grazing restores soil ecology and increases SOM and this in turn restores the soil biome. The manure the animals leave behind is food for worms, which in turn are food for birds. The longer grasses that arise in holistically grazed pastures support more insects; both the insects and the grass seeds are food for birds. The tight grazing also allows small mammals to be spotted by predatory birds.[23]

As pastures and soils become more able to soak up heavy rainfall, and release the water slowly, they reduce flooding from streams and rivers and restore watersheds and the biodiversity they support.

Complete withdrawal of nitrate and phosphate fertilisers and all pesticides – especially the persistent ones such as DDT or the neonicocide group that are thought to disrupt bee colonies – reduces dangerous pollution in soils and watercourses.[24]

The mindset of the sustainable farmer is one that wants to support biodiversity and this entails valuing wildlife on the farm. Biodynamic and organic farms are obliged to actively manage 10 per cent of the farm for biodiversity and endangered species, providing space for habitats. Biodynamic farmers in particular value bees for their pollinating activities and presence on the farm, and often provide space for top bar beehives, which provide homes for bees without a honey harvest being taken from them.

Permaculture farms and holdings have at their core the three ethics of earth care, people care and fair shares. Thus all permaculture holdings and farms will actively support biodiversity and plan to increase it right from the start of the enterprise. Diverse cropping areas, ponds and inbuilt wild areas (known as Zone Five) mean they will be naturally full of biodiversity and functional biodiversity. Permaculture systems will normally design in habitat and food supply for a range of wild flora and fauna and this is embedded in the very principles and ethos of permaculture.[25]

Sustainable food systems farmers do not perceive biodiversity as a threat to their crops. They all rely on functional biodiversity to manage pests on the farm, increase soil carbon and crop nutrition and provide preventative medicine for livestock. They actively cultivate biodiversity in corners and areas of the farm and integrate it fully into their practice.[26]

Many sustainable farm systems have short supply chains, selling direct to their customers. Thus they have a very reduced need for plastic packaging. This packaging is also actively discouraged by certification bodies. Most of the produce is sold unpackaged where possible. Plastic packaging is one of the most damaging pollutants in the wider environment. Plastic degrades into microplastics, likely to end up in the sea, where they damage marine life.

An agroecology farm supports 30 per cent more agrobiodiversity than its industrial counterpart, maintaining the important genetic and output diversity of crops and livestock.[27]

4. Producing enough food for a growing population

This is often where the argument for sustainable farming and food systems breaks down. It is documented that sustainable food systems frequently produce at least 20 per cent lower yields than the industrial systems. This is referred to as the 'yield gap'. How can we feed everyone with these systems if they have lower yields'? Are they a luxury we can't afford? The FAO suggests that 60 per cent more food will need to be produced by 2050 to feed the growing population; how can this be done by switching to systems that are undeniably good at reducing carbon emissions, are resilient to climate change and support biodiversity, but have lower yields?[28]

We can explore this problem in two ways – by focusing in on detail and by looking at an overview of the whole system. The detail of the yield data reveals that it is only in a few crops that the yield is significantly lower. To take one example: wheat that is grown in a highly industrialised system with nitrogen fertilisers and high levels of pesticides will give a higher yield than in a sustainable system. However, these yield dividends are now in decline as soils become exhausted. The yields of industrially grown grain can often be erratic. Crops can fail because of weather events driven by climate change. Both the grain varieties and the methods of working are less resilient.[29] It is economically viable to grow the lower-yielding wheat varieties, because they require no inputs.

Rebecca Laughton has demonstrated that agroecology and permaculture farms in the UK have yields of many horticultural crops higher than or equal to those of their industrial counterparts.[30] This is especially the case when one examines the yield per person on a farm, as well as the yield per unit area.[31] In the global south a switch from non-industrialised farming to agroecological systems increases yields by up to 130 per cent;[32] at worst they remain the same. If the multiple yields of all the crops from an agroecological system are added up they compare favourably with the industrial system.

Introducing agroforestry into a system increases yields by up to 40 per cent through using the vertical as well as horizontal space on the farm.[33] Holistic grazing systems restore large areas of grasslands in parts of the world where communities rely on grazing animals as their main source of food. As their grasslands degrade so does their ability to feed themselves. Across the world, holistic planned grazing is one of the few options open to pastoralists in the face of desertification.[34]

'Business as usual' in the current industrial food system is unviable. There are projections that soils in the UK are so depleted there are only 60 harvests left, or maybe as few as 40.[35] 'Land sparing' – keeping some areas in highly industrial farming systems while other areas are given over to biodiversity and carbon sequestration – is not tenable.[36] The soils that are proposed to be kept in industrial farming would have deteriorated beyond repair in 40–60 years and would have to switch to a regenerative form of farming in the long term.

If we approach the problem from the other perspective, by examining the whole food system, we find that sustainable food systems can feed the world. We can ask ourselves, 'What if small-scale local sustainable food production

models were replicated across the world?' What might that look like? Many researchers have done this and have found in their models that sustainable food systems can feed the increasing world population with healthy, affordable food.[37]

How can this be done? There are two main strategies: by reducing food waste, and by changing to a more plant-based diet. Currently, 30 per cent of food produced is wasted. By reducing the long supply chains, or localising food, a great deal of waste can be reduced.[38] Currently, 40 per cent of grain is fed to livestock or used for biofuels. If more people adopted a local and plant-based diet, this food can be eaten by people rather than livestock; one researcher suggests that this switch can produce 70 per cent more food for human consumption. The by-product of this is a healthier diet.

In practice these models draw extensively on sustainable food practice: making equitable and healthy food available to everyone, in economically sustainable ways for both the farmer and the consumer.

How can enough food be produced from sustainable food systems?

1. Reducing food waste – shorter supply chains

In large-scale industrial food systems there is a huge amount of waste. The FAO estimates that 30 per cent of all food produced is wasted in the supply chain. There is vast inequality of food distribution in the world, starvation happening in many places in the world while there is an epidemic of obesity in the global north. Enough food is produced for all but it is shared in an inequitable way. The waste occurs on the farm, with seasonal gluts, in long transport systems, in grade-outs, in the shops and in the home. With shorter supply chains the amount of food waste drops because there is a more secure route to the consumer with only one or two transactional steps.[40] Food is delivered in a fresher condition and has a longer shelf life.

Customers are happy with wonky vegetables and fruit. Small-scale producers often have food processing hubs where gluts can be processed, and livestock that can be fed the grade-outs.

During the 1970s and 1980s, as the industrialisation of food took hold in the global north, local food systems disappeared with the advent of refrigeration, supermarkets and highly processed and commodified food. This led to the implementation of the economies of scale on farms, and they became, and are still becoming, ever larger and more specialised. To counteract this, the sustainable farming systems have since the 1980s pioneered methods of shortening the supply chain, selling directly to people who are happy to have seasonal, unpackaged and misshapen produce.

These direct trading systems allow the farmer or grower to increase their revenue: the producer receives 100 per cent of the retail value of the food, in contrast to the industrial long supply chain model in which the farmer often receives only seven per cent of the value of the product, the rest going to the supermarkets and wholesalers.[41]

The direct trading systems that have been developed deal primarily with selling fruit and vegetables, milk and meat, and not with the crops that are grown on a larger scale and are the most commodified forms of food – such as grains

and pulses. The latter are rarely sold via direct marketing systems. However, companies such as Hodmedods are creating shorter supply chains for grains and pulses. They have created online marketplaces to sell grains and pulses that have been grown in the UK by named farmers, thereby connecting the farmer and the consumer. They buy directly from the farmers and sell directly to the consumer or to shops, shortening the supply chain to one or two transactions. The UK Grain Network links farmers and bakers to allow direct sales between them. Similar systems are also being developed in the US.[42]

Through direct trading, food miles and packaging are reduced and the food is fresher, it lasts longer and so less is wasted. Trust and knowledge are developed between the food producer and the consumer, bringing about a vibrant, healthy local food culture. The most common forms of direct trading systems are CSA and farmers' markets.

Community-supported agriculture:
The concept of CSA, sometimes called 'subscription farming', arose from biodynamic farming in the US in the 1980s. Trauger Groh realised that, as a farmer, he was being pushed into reducing the number of crops he grew as he sold his produce increasingly into the supermarket sector. This meant that he took all of the economic risk in producing food. At the same time the local consumers near his farm were keen to buy and eat a varied range of biodynamic food, but this was not available in the supermarkets. The threefold philosophy of biodynamic farming – integrating the social, economic and environmental spheres – inspired him to dream up the concept of CSA.

If consumers wanted a farmer to grow and sell seasonal, healthy fresh food all year round then they would need to enter into an economic agreement with the farmer to share the economic risk of food production and share any gluts from the farm.[43]

Every growing season some crops do better than others, creating gluts of some crops and scarcity of others. The skill of the farmer and the choice of land and microclimate can bring the occasional financial benefit from producing the earliest strawberries or having carrots when no one else does. However, in most cases all of the farmers and growers in a region will have the same gluts and crop failures, owing to prevailing weather conditions. If a farmer has managed to have a magnificent crop of strawberries, it is likely that everyone else in the region will have done the same, driving the price down. As supply chains consolidated to supply supermarkets in the 1980s, pressure was put on farmers and growers to reduce their product lines, to specialise, and so exploit the economies scale of a few crops, which increased the economic risk if there was a crop failure because of bad weather. This trend also meant that anyone wanting to eat a diet of organic or biodynamic food struggled to find a complete range of food in a supermarket.

The biodynamic farmers in New York State had a system rethink. Those of them who created the concept of CSA asked their customers to buy their crops up front with a subscription. They worked out how much money they needed to support themselves and run their farms, and how many customers they could feed. They charged a fee that covered the whole of the following year's

Figure 9.2: CSA bags ready to go out, Huxhams Cross Farm.

crops. These pioneers grew a range of crops and then shared the harvest with the customers. If there was a glut of strawberries the customers shared the glut. If there were no carrots they had no carrots in their weekly delivery. The farmer received a regular income; the customer received better value for their money and access to the farm and a range of organic or biodynamic food. The concept of the box scheme arose out of this system, whereby a customer pays a certain amount of money and receives a regular box of seasonal vegetables or fruit for six months or a year. In the original thinking, the community also took part in some of the farm work, during the potato harvest or regular volunteer days, to ease the work pressure on the farmer and also to experience farm life.

Box schemes and CSAs have now taken off across the US and Europe, one of the largest operators being Riverford Farm in the UK, which delivers more than 50,000 boxes per week. There is a spectrum of local CSA models in operation. At one end of the spectrum is the subscription model, where the customer makes a commitment to buy produce for a season and contributes to work on the farm. At the other end of the spectrum a customer can order on a weekly basis and does not need to be active in farm life. CSAs have been set up for fruit and vegetables, meat, eggs, milk and cut flowers (see Figure 9.2).

Farmers' markets and farmers' assemblies:
Markets have been in existence for millennia; they are ancient trading places, and traditionally farmers went to sell their produce on a weekly basis and catch up with the neighbours. Most people love market shopping, since it is sociable and fun. In Europe and the US, supermarkets supply over 90 per cent of all food sales and the farmer generally has no knowledge of or contact with the consumers; the farmer no longer goes to market to sell their produce. The farmers' markets that arose in the 1990s reconnected the farmer to the consumer (Figure 9.3).

Farmers gather weekly or monthly to sell their produce directly to their customers, which allows relationships to build up between the farmers, growers and consumers and means that 100 per cent of the value of the produce goes to the producers. The farmers pay a fee for their pitch and the market coordinators do the marketing for the farmers.

A 'farmers' assembly' is an online version of a farmers' market. Food producers list the produce they have available and it is ordered online; the farmers then drop it off at a given place and time. A team collates the orders and the customers come and collect their collated order. This is less time consuming for the farmer but there is less contact between the farmer and the consumer. The online Open Food Network lists virtual markets and producers around the UK, which facilitates the joining up of producers and consumers across the country.

Similarly Reko organises local food delivery via Facebook across Europe.

Relocalising and shortening food supply chains regenerates the food economy and builds resilient businesses. It is sociable, builds community, supplies healthy food to local people and changes eating habits.

2. Changing diets

Many studies have shown that feeding a growing population from sustainable food systems is possible if people reduce their meat consumption and eat a more plant-based diet. Currently, 40

Figure 9.3: Farmers' market, Growing Communities, Hackney, London.

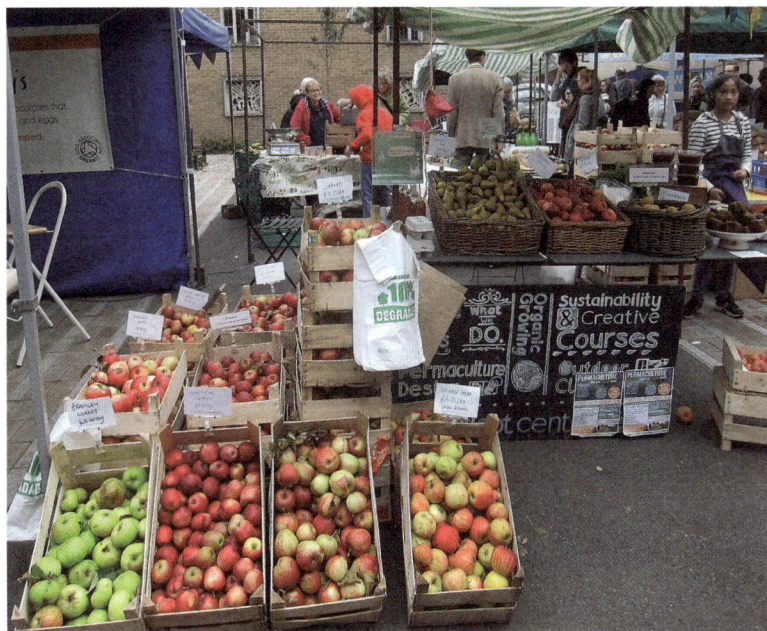

Growing Communities, London

Growing Communities set up their farmers' market in 2000 as a way of accessing local low carbon healthy food in London. Julie Brown and Kerry Rankine invited local producers from within a 95 km radius of London. They chose organic status as a way of defining what 'low carbon' meant. The market is held weekly in Hackney in London and offers a range of vegetables, fruit, meat, fish, processed food – such as fermented food – mushrooms, milk and cheese. Tea and coffee and snacks are sold, with places to sit, and regular musical spots and festive events such as apple days help to bring a good crowd every week. The street food stalls on the market are obliged to source 70 per cent of their ingredients from the stallholders.

Growing Communities also run a form of CSA by sourcing wholesale produce from the farmers who sell at the weekly farmers' market, and others, running a weekly vegetable round to supply more than 1,000 customers per week. They have also created a 'patchwork farm' of two hectares of food production in Hackney itself, made up of a number of small patches of land. On these they grow salad all year round in intensive beds and polytunnels.

Growing Communities are now replicating this model in other cities with the Better FoodShed concept. The aim is to recreate sustainable food supply chains into cities, while creating economically viable markets for new sustainable farms and growers in and around cities.

After many years of trading, observation and analysis of the figures, Brown came up with the concept of Food Zones to show the percentage of the food sold in the weekly farmers' markets and through CSA, and how far it has travelled. These figures are derived from years of sourcing the supply of fruit and vegetables into the weekly CSA run by Growing Communities (Figure 9.4). Growing Communities' strapline is 'Changing the world one carrot at a time'.[44]

Figure 9.4: The food zones.

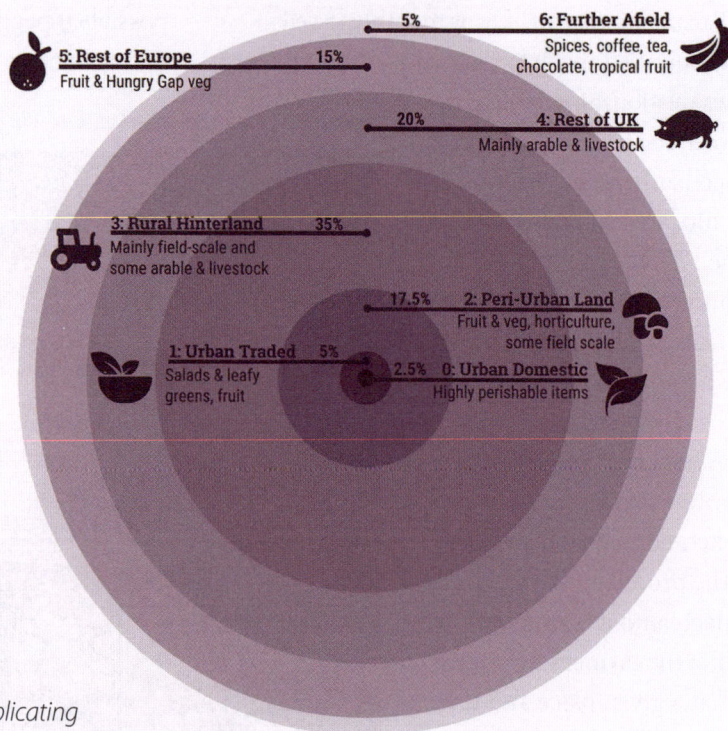

Zone	Description	%
6: Further Afield	Spices, coffee, tea, chocolate, tropical fruit	5%
5: Rest of Europe	Fruit & Hungry Gap veg	15%
4: Rest of UK	Mainly arable & livestock	20%
3: Rural Hinterland	Mainly field-scale and some arable & livestock	35%
2: Peri-Urban Land	Fruit & veg, horticulture, some field scale	17.5%
1: Urban Traded	Salads & leafy greens, fruit	5%
0: Urban Domestic	Highly perishable items	2.5%

per cent of grain crops grown are fed to animals or used in biofuels.[45]

Red meat consumption is linked to cancer, and cancer research suggests that 20–30kg of meat per year per person is a healthy amount. In Europe, the US and Australia the per capita consumption of meat per year is 80–110kg. Reducing the amount of meat eaten would not only be healthy but lead to a more sustainable diet. The 'nutrition transition' that has followed the industrialism of agriculture has increased demand for meat in many countries. In some cases the industrialisation of beef cattle has created giant feedlots where cattle are fed on grain imports from across the globe, the resulting beef is exported across the globe and they produce methane that contributes to climate change.

The EAT-Lancet report suggested a shift to a planetary healthy diet that would consist of 'half a plate of fruits, vegetables and nuts. The other half consists of primarily whole grains, plant proteins (beans, lentils, pulses), unsaturated plant oils, modest amounts of meat and dairy, and some added sugars and starchy vegetables.' A planetary healthy diet is one that is good for health and good for the environment and can feed everyone from the planet's resources.[46]

Sustainable food systems create small mixed farms that produce the ingredients for this 'planetary sustainable' diet. It turns out that not only is it more sustainable but it is also delicious and has the extra dividend of being healthy. Colin Tudge refers to the 'gourmet' diet that arises from farms that are based on a

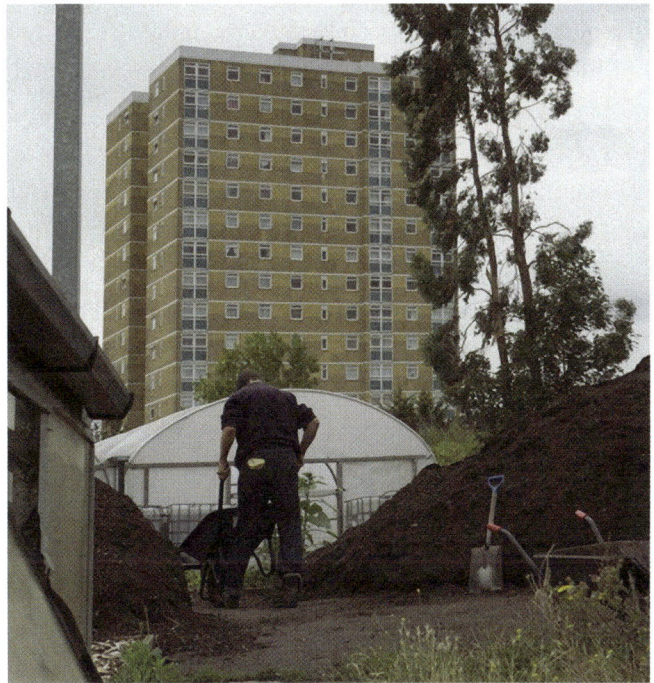

Figure 9.5: Dagenham Farm, Growing Communities.

closed loop system and the carrying capacity of their site. A small-scale mixed farm with a closed loop system grows lots of vegetables and fruit, and some grains and pulses, and has a few cows or sheep on grassland, and poultry in mobile houses, to eat the waste crops, graze the green manures and tidy up at the end of a crop. In Italy a typical gourmet meal from a small mixed, sustainable farm might comprise pasta and bolognese sauce with olives, cheese, salad and wine; in China it might be rice, vegetables, duck and plum sauce, and tea.[47]

Healthy food and healthy people

Not only do biodynamic, organic, permaculture, agroforestry, agroecology and regenerative food systems meet the four challenges, but they also provide another yield – that of health.

The global north is facing a health epidemic of illness caused by poor nutrition, and much of that is from eating food produced in the industrial farming system. Large-scale changes to sustainable food systems will address these root causes of ill health.[48]

Eve Balfour said, 'If the nation's health depends on the way its food is grown, then agriculture must be looked upon as one of the health services, in fact, the primary health service.'[49] It is assumed that foods grown from sustainable systems are more healthy, but how does this food differ from the food produced in the industrial farming system and how does it affect the health and well-being of the people who are eating it? What makes the food from a sustainable system more healthy than food from an industrial system?

The farmer and poet Wendell Berry said that 'eating is an agricultural act'. What he meant is that the choices of which food we buy, from whom and how we prepare it, cook it and eat it make up our 'food culture'. This food culture can have a dramatic effect on our health and well-being.[50]

Sustainable food systems tend to be relocalised, seasonal and are based on the concept of the 'foodshed' – that is, diverse food that is produced locally within a bioregion and sold to local people via something like a CSA or farmers' markets. It has been shown that this approach to buying and selling food increases the consumer's consumption of fresh fruit and vegetables and improves the financial security of the farmer. Local food cultures often celebrate foods and the seasonality of food. This might be the forgotten festival of Lammas in Christian countries to celebrate the first wheat harvest with

a celebratory loaf and party, or the Moon Festival in China to celebrate the harvest with mooncakes and a thanksgiving party. These celebrations are times for a pause in farm work, help to build communities, link farmers with the consumers of their produce and increase well-being all round.[51]

The food culture around shopping for local seasonal food increases the range of diet and changes eating habits. We all know that a healthy diet should include at least five portions of fresh fruit and vegetables a day. Tim Spector has suggested that we should eat up to 20–30 'biologically distinct plant foods' per week in order to stay healthy. 'Biodiversity needs to be on a consumer's plate, routinely,'[52] or, as Tudge says, 'plenty of plants, not much meat and lots of variety'.[53]

How does increasing the range of fruit and vegetables we eat improve our health? If fruit and vegetables are from sustainable farms, they will be 'nutrient dense', meaning they have higher levels of minerals and secondary metabolites in them, which is what gives them their taste. This means that when we eat them we obtain higher levels of nutrition, minerals and what are often called 'secondary metabolites'. The higher levels of minerals and micronutrients in the food are thought to arise from the rich soil biome they are grown in, which facilitates increased absorption of minerals from the soil. The higher levels of minerals, micronutrients and vitamins found in the food literally 'feed' people better, bringing about better digestion and better mineral and vitamin absorption by the body.[54]

A second mechanism is that eating a wide range of fruit and vegetables supports a healthy gut biome. Vegetables from sustainable systems

Impact of pesticides on human beings

DDT is an organochlorine insecticide banned in the US in 1972 but still in use in some parts of the world today. It has long-term persistence and has accumulated in the food chain, particularly in the Arctic. It has found its way, in high levels, into the breast milk of the Inuit peoples, among whom it has caused and still causes the suppression of the immune system (immunotoxicity) of babies, giving them very high rates of infectious diseases. Because of this, Inuit women are recommended to bottle feed their babies. DDT residues are found in such high levels in some pastures in New Zealand that they cannot be grazed with cattle.

Glyphosate and plastics are known to cause endocrine disruption. This causes the disruption of hormones by mimicking oestrogen. Glyphosate, or Roundup, is known to be routinely sprayed on wheat as a desiccant just before harvest. In 2015 the World Health Organization's International Agency for Research on Cancer (IARC), concluded that 'Glyphosate is probably carcinogenic to humans'.[60] In the UK, DEFRA's routine testing of bread found that 30 per cent contained traces of glyphosate.[61] Endocrine disruptors cause cancer and impact on fertility levels in both men and women.[62]

often have traces of soil left on them which are rich with soil microflora, and eating small amounts of soil increases gut biome health. Experiments show that when mice ingest the soil bacterium *Mycobacterium vaccae*, neurons in their brain produce more serotonin. The gut biome is also improved by eating fermented foods and probiotic drinks and minimising the use of antibiotics.[55]

Recent research on the gut biome suggests that the human gut can absorb more nutrients and secondary metabolites when the gut biome is healthy. Just as the soil feeds the plant if it is full of microorganisms that facilitate the plant's uptake of minerals from the soil, the gut biome facilitates the uptake of nutrients and secondary metabolites from the food that is eaten and then digested.

Further recent research has demonstrated links between a healthy gut biome, an improved immune system, the ability to deal with stress and anxiety, and improved mental health generally. The immune system and the brain are intimately connected; if the immune system is not working well it can cause illness associated with prolonged inflammation such as arthritis and asthma, and this is now also linked to periods of depression and other mental health conditions.[56]

Sustainable food systems suggest an approach to health in which it is recognised that, instead of just treating illnesses, it is preferable to promote wellness, through good nutrition as a form of preventive medicine. The Pioneer Health Foundation tried to do just that when it set up the Pioneer Health Centre, an initiative also commonly known as the 'Peckham Experiment', which ran from 1926 until 1950. The Pioneer Health Centre was set up to promote wellness rather than the curing of disease and was, at its core, based on creating a positive food culture. The team of doctors who set up and ran the centre invited local families to join it. The families were given an annual consultation, at which they were given information on how to become more healthy. The centre had an organic farm, where the families were encouraged to grow food. This food ended up in the canteen. There was also a gym and a swimming pool, and community events were organised. The founders of the Pioneer Health Centre were also key figures in the early organic movement. They were keen that the British NHS should be set up not to cure disease but to promote wellness. They lost the argument.[57]

The final, simple reason why sustainably grown food is healthier is the absence of pesticides or herbicides, nitrogen fertilisers and antibiotics, since all these things damage human health. Antibiotics harm gut biome health. People regularly consume antibiotics when they eat large amounts of processed meat, eggs or milk from industrial livestock systems, since these contain high levels of antibiotics. Smaller amounts of unprocessed meat, eggs and milk that do not contain antibiotics are healthy to eat. Antibiotics are used in industrial farming as a prophylactic to prevent outbreaks of disease caused by high stocking density.

They also act as a growth stimulant for animals, resulting in greater weight and therefore higher profits. Not only do the antibiotics damage our gut biome; overuse of them also raises the chance of bacterial diseases and infections becoming immune to antibiotics. The British Medical Association (BMA) has warned, 'The risk to human health from antibiotic resistance developing in micro-organisms is one of the major public health threats that will be faced in the 21st century.'[58]

Pesticides currently used in farming are cleared as safe, but different countries have different standards. It has been found time and time again that human health issues arise long after pesticides have entered a food chain, that damage does ensue from long-term, low-level exposure, albeit that the majority of health problems from pesticides arise in people with high levels of exposure – the farmers or growers in countries where pesticides are overused and contaminate the environment. The risk from eating food that carries pesticide residue or cocktails of pesticide residue is low, but it is also an unknown risk. The Soil Association report on 'Organic Farming, Food Quality and Human Health' reviews the evidence thoroughly; it summarises that long-term exposure to pesticides residues in food is known to cause neurotoxicity, endocrine disruption and immunotoxicity and that pesticides are also carcinogens. High levels of nitrogen fertilisers, especially on leafy vegetables, are also proven to be carcinogenic; levels are by law kept low on green vegetables in Europe for that reason.[59] ∎

In a nutshell

All of the sustainable farming systems:

- have a lower carbon footprint than industrial systems and sequester more carbon in their soils and greater tree cover;

- are more able to adapt to climate change because of their greater SOM, frequent mulching and diverse cropping systems;

- support more biodiversity because of the greater SOM and soil biodiversity, as well as having more habitats on the farms;

- are able to produce enough food to feed increasing populations, by reducing food waste via shorter food supply chains and selling directly to the customer;

- encourage people to change to a more localised and plant-based diet;

- provide a healthier diet;

- create a circular food economy;

- improve people's health because food from sustainable systems has higher nutrient density and more genetic repair enzymes, improves the gut biome and does not contain harmful additives and residues such as pesticides, antibiotics and high nitrate levels.

1 Attributed to Albert Einstein.

2 D. Holmgren, *Permaculture Principles and Pathways*, Permanent, East Meon, UK, 2011.

3 T.F. Stocker, D. Qin, G.K. Plattner, M. Tignor, S.K. Allen, J. Boschung, A. Nauels, Y. Xia, V. Bex and P.M. Midgley (eds), *Climate Change 2013: The Physical Science Basis*, Cambridge University Press, Cambridge, 2013.

4 https://www.savory.global

5 H. Willer, B. Schlatter, J. Trávníček, L. Kemper and A. Lernoud, 'The World of Organic Agriculture Statistics and Emerging Trends', FiBL/IFOAM. 2020.

6 https://www.agroforestryresearchtrust.co.uk

7 T. Groh and S. McFadden, *Farms of Tomorrow*, Biodynamic Farming and Gardening Association, Kimberton, PA, 1990.

8 J.M. Lavelle et al., 'Conceptualizing Soil Organic Matter into Particulate and Mineral-Associated Forms to Address Global Change in the 21st Century', *Global Change Biology*, 2019.

9 Soil Association, 'Soil Carbon and Organic Farming', 2009.

10 J. Moyer, A. Smith, Y. Rui and J. Hayden, 'Regenerative Agriculture and the Soil Carbon Solution', Rodale Soil Carbon White Paper, 2020.

11 Soil Association, *The Agroforestry Handbook*, Bristol, 2019; J. Pretty, *Agri-Culture*, Earthscan, London, 2002; https://yatesbury.wixsite.com/yatesbury/impact

12 Soil Association, *The Agroforestry Handbook*.

13 Moyer et al., 'Regenerative Agriculture and the Soil Carbon Solution'.

14 https://www.savory.global

15 J.M. Lavelle et al., 'Conceptualizing Soil Organic Matter into Particulate and Mineral-Associated Forms to Address Global Change in the 21st Century', *Global Change Biology*, 2019; P. Cortrufo, 'Simple Biophysics of Soil Carbon Sequestration', Global Carbon Management Workshop, 23 September 2020

16 https://www.savory.global; Willer et al., 'The World of Organic Agriculture Statistics and Emerging Trends'.

17 Soil Association, *The Agroforestry Handbook*.

18 Groh and McFadden, *Farms of Tomorrow*.

19 C. Massy, *Call of the Reed Warbler*, Chelsea Green, Hartford, VT, 2017.

20 L. Woodward, J. Smith, N. Pearce, M. Wolfe and N. Lampkin, 'The Biodiversity Benefits of Organic Farming', 2010.

21 https://www.thephf.org/peckhamexperiment

22 Soil Association, *The Agroforestry Handbook*.

23 P. Mader, A. Fliessbach, D. Dubois, L. Gunst, P. Fried and U. Niggli, 'Soil Fertility and Biodiversity in Organic Farming', *Science* 296, 2002, pp. 1694–1697.

24 Woodward et al., 'The Biodiversity Benefits of Organic Farming'.

25 https://www.pastureforlife.org

26 Woodward et al., 'The Biodiversity Benefits of Organic Farming'.

27 Pretty, *Agri-Culture*.

28 B. Mollison, *Permaculture: A Designer's Manual*, Tagari, Tyalgum, 1988.

29 https://www.thephf.org/peckhamexperiment

30 R. Laughton, 'A Matter of Scale: A Study of Productivity, Financial Viability and Multifunctional Benefits of Small Farms 20 hectares and Less', Landworkers' Alliance, 2017.

31 FAO, 'Climate Change, Agriculture and Food Security', 2016.

32 Laughton, 'A Matter of Scale'.

33 Soil Association, *The Agroforestry Handbook*.

34 https://www.savory.global

35 Mollison, *Permaculture*.

36 http://www.fao.org/organicag/oa-faq/oa-faq1/en/

37 FAO, 'Can Organic Farmers Produce Enough Food for Everyone?'; L. Woodward, 'Can Organic Farming Feed the World?'. Elm Farm Research Centre, 1995; A. Muller et al., 'Strategies for Feeding the World More Sustainably with Organic Agriculture', *Nature Communication* 8(1290), 2017; J. Pretty et al., *Assessment of the Growth of Social Groups for Sustainable Agriculture and Land Management*, Cambridge University Press, Cambridge, 2020.

38 P. Holden, 'Landsparing vs Landsharing', Sustainable Food Trust, 2020.

39 FAO, 'The State of Food and Agriculture: Moving Forward on Food Loss and Waste Reduction', 2019. 'Redefining Agricultural Yields: From Tonnes to People Nourished per Hectare 2013', *Environmental Research Letters 8, 2013, pp. 1–8* doi:10.1088/1748-9326/8/3/034015

40 Mader et al., 'Soil Fertility and Biodiversity in Organic Farming'.

41 https://www.fao.org/organicag/oa-faq

42 T. Lang and M. Heasman, *Food Wars*, Earthscan, London, 2004.

43 Groh and McFadden, *Farms of Tomorrow*.

44 http://www.smallfoodbakery.com/yq-wakelyns-population

45 Moyer et al., 'Regenerative Agriculture and the Soil Carbon Solution'.

46 EAT-Lancet Commission, 'The EAT-Lancet Report', 2019.

47 C. Tudge, *Good Food for Everyone Forever*, Pari, Pari, Italy, 2011.

48 Tudge, *Good Food for Everyone Forever*.

49 Soil Association, 'Organic Farming Food Quality and Human Health', 2001.

50 W. Berry, *What Are People For?*, Random Century, London, 1990.

51 Tudge, *Good Food for Everyone Forever*.

52 T. Spector, *The Diet Myth*, Weidenfeld & Nicolson, London, 2016.

53 Tudge, *Good Food for Everyone Forever*.

54 A.M. Mayer, 'Historical Changes in the Mineral Content of Fruits and Vegetables', *British Food Journal* 99(6), 1997, pp. 207–11.

55 M. Butler, S. Morkl, K. Sandhu, J. Cyran and T. Dinan, 'The Gut Microbiome and Mental Health: What Should We Tell Our Patients?', *Canadian Journal of Psychiatry* 64(11), 2019, pp. 747–760.

56 Ibid.

57 https://thephf.org/peckhamexperiment

58 British Medical Association, 'Antimicrobial Resistance: Ambition to Action', 2019.

59 Soil Association, 'Organic Farming Food Quality and Human Health'.

60 IARC, *Some Organophosphate Insecticides and Herbicides*, Lyons, 2015.

61 Pesticide Residues in Food (PRIF), 'Annual Report 2018'.

62 S.H. Swann and S. Colino, *Countdown: How Our Modern World Is Threatening Sperm Counts*, Simon & Schuster, New York, 2021.

Chapter 10

Designing the world we want

Buy land, they're not making it anymore.[1]
Mark Twain

Introduction to a case study: the Apricot Centre @ Huxhams Cross Farm

The Apricot Centre team in partnership with the Biodynamic Land Trust came up with an answer to 'designing the world we want',[2] and this has been the creation of Huxhams Cross Farm. This is the story of the creation of this farm.

In 2015 the Biodynamic Land Trust bought Huxhams Cross Farm in Dartington, near Totnes, in Devon, with investment from 150 shareholders. In reality the farm was little more than a collection of six degraded fields of 13 hectares with no farm buildings. It had been farmed industrially for the last 40–50 years by the main tenant of Dartington Hall, a dairy farmer, with three arable fields of continuous barley, two wet meadows that had been abandoned and one field that had been put into set-aside and sprayed with glyphosate for many years. The soil structure was so damaged the land was just a giant muddy puddle that could barely grow grass. The contractor called it 'a miserable bit of land'.

The story of this small farm reflects the journey of many farms throughout the UK and the world. In the 1800s it was owned by Henry Champernowne, the owner of Dartington Hall. It was rented out to three tenants, and the old maps show that they grew vegetables, fruit in an orchard, and arable crops, with some livestock on the meadows. The barn on the farm was unusual for Devon, in that it comprised a threshing barn upstairs, for processing the wheat crops, and a cow shed downstairs; it was built next to the springs. There was a 'great meadow' for grazing draft animals. The apple barn was next door, and the cider press just around the corner. The parcel of land that we now call Huxhams Cross Farm was sold off to small-scale farmers at the turn of the 20th century, the time of the farming depression, and farmed on a small scale with arable crops until it was sold back to Dartington Hall Trust in the 1930s.

When Dorothy and Leonard Elmhurst bought Dartington Hall Estate in 1920, the 14th-century Dartington Hall was derelict and the farm run down. Dorothy, an American, was one of the richest women in the world. She and her English husband regenerated the estate as a place to live and work and experiment with progressive arts and farming. Dartington Hall became a progressive centre for arts, crafts, education, architecture and thinkers. It was here that the concept of the NHS was born. However, when it came to farming, the 'progressive of the day' was the new industrial model of farming. Leonard Elmhurst had studied agriculture at Cornell University, and so Dartington Hall Trust pioneered industrial agriculture on the estate. Huxhams Cross Farm was bought to enlarge the scale of the dairy farm. A farmer was brought over from Denmark to modernise the dairy farm. He pulled out old hedgerows and Devon banks[3] and introduced tractors, fertilisers and pesticides. The estate experimented with artificial insemination of cattle and battery farming of chickens. This continued with the subsequent tenants and the estate farmed industrially right up until 2015.

Over the years, Dartington Hall was home for a while to some of the founders of the organic movement. Eve Balfour visited. John Seymour visited in the 1970s and made recomendations for the whole of Dartington Hall Estate to be converted into an organic farm. Lawrence Woodward, the founder of the Organic Research Centre, was educated at the progressive school on the estate. Schumacher College started on the estate in 1991, and many of the most prominent environmental thinkers visited and taught on the estate. I personally set up what is now School Farm, an organic market garden, in 1989, at the suggestion of David Cadman and in response to a conference held by Wendell Berry on the estate. School Farm CSA and the Schumacher College gardens were the only organic food producers on the estate until 2020, when Old Parsonage Farm went into registered organic conversion, This farm run by Jon and Lynne Perkin practises agroforestry, grows a mix of population wheat and landrace wheat, hemp and other ancient

grains, raises pasture-fed cattle and is a partner in the Dartington Mill CIC (community interest company).

Huxhams Cross Farm, although part of the dairy farm, was sown with continuous barley as a commodity crop; its stubble fields were left bare every winter, the wetland meadows were abandoned and the great meadow was sprayed off with glyphosate to kill the 'weeds' and grass. This was subsidised as a form of 'set-aside', the removal of farm land from food production in Europe in the late 20th century because of the overproduction of food. One of the farm workers commited suicide in the barn and it was subsequently abandoned and then sold off to be a holiday home when the farm was sold to the Biodynamic Land Trust.

This tiny farm's story is like that of many, and yet it occupies a unique position; it is part of a thriving local culture of small- and large-scale organic, biodynamic, permaculture, agroecology and agroforestry farms and holdings and rewilding projects in and around Totnes. Huxhams Cross Farm is now registered biodynamic, has been designed using permaculture methodology and weaves in agroforestry methods throughout. We used the toolkit of different farming systems to create a regenerative farm.

The Biodynamic Land Trust's mission is to secure farms into long-term trusteeship for sustainable food production for the sake of farmers and communities.[4] Dartington Hall Trust wanted to implement a 'land partnership' scheme with many smaller tenants on its estate who would practise different forms of sustainable farming to create a dynamic food culture, as well as establish a world class learning campus for sustainable agriculture next door to Schumacher College. Huxhams Cross Farm was the biodynamic farm in the mix.

Farmland in the UK more than doubled in price between 2010 and 2015, fetching £19,000 to £30,000 a hectare and putting the price of the average small-sized farm with a farmhouse and buildings out of reach of most people. The capital required to buy a farm does not make financial sense, since the mortgage payments that need to be met will far outstrip any income that can be generated from farming, especially with the long-term investment and time required to build sustainable systems of food production. The Biodynamic Land Trust's solution to this is to buy 10- to 20-hectare plots made up of five to seven fields, costing somewhere between £200,000 and £370,000. In order to make a living on this size of farm, there has to be a mix of high-value horticultural crops, small-scale agriculture producing high-value produce such as organic eggs and grass-fed meat, and value-added products. All these then need to be sold direct to the end customer, and appropriate non-food diversification may also be required in order to make a viable enterprise. Such diversification can sometimes account for up to 70–90 per cent of farm income.

The Biodynamic Land Trust has pioneered the 'farm community buyout' method using a layered cake of 'community shares', gifts and loans. It is a 'community benefit society' and has charitable status. To buy a farm, the Biodynamic Land Trust offers withdrawable shares to individuals, both in the farm's local area and nationally.

According to the warmth of local support and fundraising effort, the Biodynamic Land Trust has managed to raise between 30 and 75 per cent of the sums required to fund three small farm purchases, the rest of the money coming from interest-free loans and from an endowment. Shares deliver no financial reward, only the knowledge that you are a 'trustee', or shareholder, of a farm that brings environmental and social benefits to the locality. Initially, it was thought that mainly local people would invest in local farms, but it turns out that national and international investors like the idea of owning a bit of a farm somewhere in the UK if the vision is clear enough. The strapline of the Biodynamic Land Trust is 'Changing the world one farm at a time'.[5]

The permaculture design of the farm

Survey process

Once Huxhams Cross Farm had been bought, the team needed to make friends with the land and get to know it. So we walked, poked around the corners, took soil samples and tested them for phosphate, potassium, the pH and organic matter. We sat in the wind, watched the sun's movements, followed water down the slopes and as it popped out on the keylines. We measured altitudes and slopes and did contour mapping. We looked at old maps, and gazed at hedges, admired views and stood in howling gales in the rain to see where the wind came from. We followed deer tracks and talked to the dog walkers and neighbours to find out what was and what had been possible on the site. We spoke to the previous contractors and farmers who had worked the soil. We also sat down for some quiet observation time, to get a 'sense of place', or of the dreaming of the place, the *genius loci*.

We spoke to the stakeholders: the neighbours, the biodynamic community, the permaculture community, the science-based sustainable agriculture community, the people interested in local, good-quality food. We spoke to Dartington Hall Trust, who were selling the land, and to other local food producers to find out how we could collaborate. We spoke to many people who didn't like what we were doing. We spoke to the national leadership of the Permaculture Association and the Biodynamic Association and asked them what they would like to see happen on the farm. We spoke at length to the Biodynamic Land Trust about their wishes for the farm. The Apricot Centre team spent time deciding what our vision and skills were and how far we were going to stretch them, how much money we needed to earn to support ourselves and how much money and time we could invest in the development phase. We used online survey monkeys, paper-based surveys, conversation and dialogue to engage with a wide range of people in this consultation process.

Analysis and design process

The Biodynamic Land Trust suggested that we do the core of the design work for Huxhams Cross Farm in a workshop format so that other people could see how it was done. So we organised two weekends where we analysed all of the information and designed first the pattern and then the detail of the emerging new farm. Both were wonderful events, involving small numbers of people who worked incredibly hard to grapple with, play with and create the concept of what is now Huxhams Cross Farm.

Function and element analysis

From the results of the survey work we created a list of the functions that this particular farm should fulfil:

- produce good quality food – vegetables, fruit, eggs and some grain
- support biodiversity
- sequester carbon
- resilience to climate change
- offer access to children and community
- offer a wellbeing service
- be a demonstration farm
- carry out research ansd training
- be economically viable
- be beautiful.

From this we also had a list of the 'things', or

Figure 10.1: Function element analysis for the permaculture design at Huxhams.

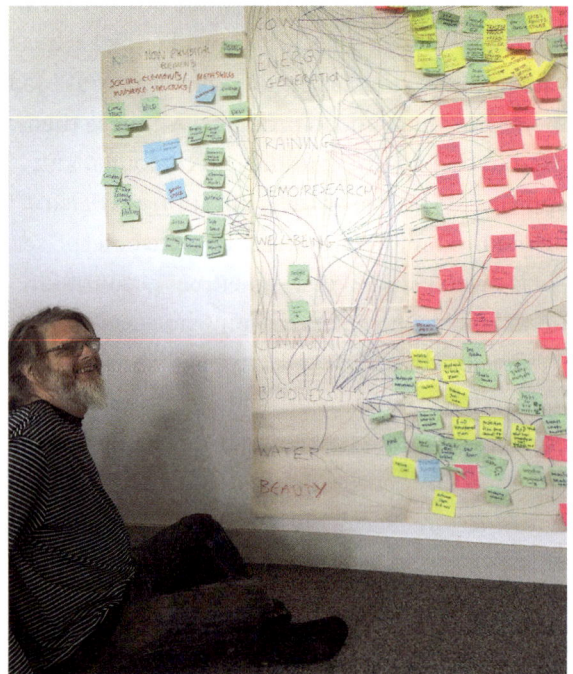

in permaculture jargon 'elements', that the farm would contain in order to fulfil the above functions, such as polytunnels, orchards, chicken houses, forest school area, farm trail, training room, barn, toilets, seats to enjoy the view, and so on. It was a very long list and we have not included it here.

We then carried out function and element analysis, that is, making sure that each function is supported by more than one element and each element supports more than one function in order to sustain resilient systems (Figure 10.1).

Zone and sector analysis

On large-scale maps of the site we mapped out on overlays the direction of the wind, the flow of the water, the sunny and shady spots (Figure 10.2). On another overlay we mapped out the flow of people on the farm: where we would have a cup of tea, what areas we would go to most often and why, and where the paths would be.

Designing

We made a huge scale map of the farm using a large sheet and gaffer tape, put it on the floor and then got out toys, plasticine and twigs, cardboard and scissors. The design group then spent a wonderful few hours modelling what the new farm might look like. We placed the elements in the right zone and in the right sector to make the most of the farm workers' time and energy. The vegetable-growing area was placed in or around where the barn and training centre would be. The orchards and arable or pasture fields were placed furthest from the training centre, since they needed to be visited less regularly. The site had some constraints.

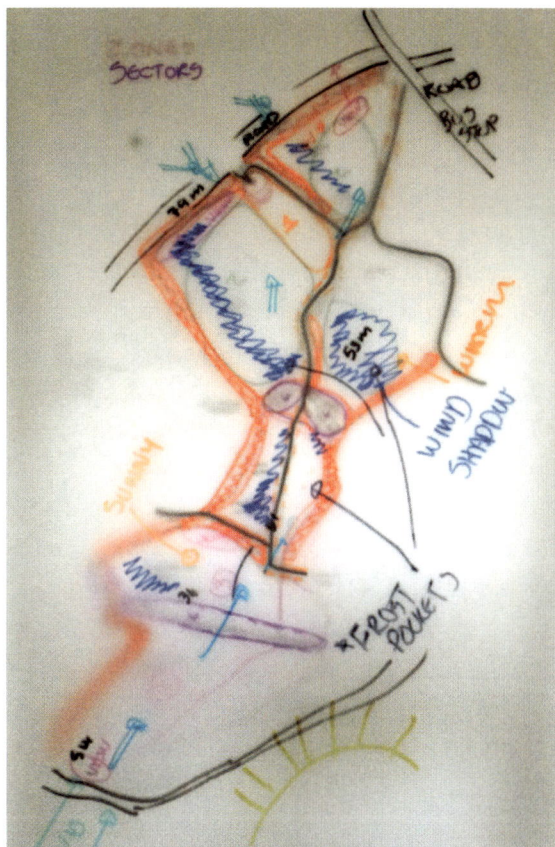

Figure 5.4: Sector analysis.

There was only one place where the farm could be accessed by car, so the buildings, that is, the barn and the training centre, had to go next to that entrance, since the cost of long tracks through the farm was prohibitive. The centre of the farm is one of two wetland meadows of huge conservation and local value, with an ancient droving track running through it, still used by many walkers. This meadow is full of orchids, insects, owls and bats. Its use cannot be changed and so we decided to improve the biodiversity here and on the second wetland meadow. To this end, two Shetland cattle were put in the meadows as conservation grazers. These cattle

Figure 10.3. Final design drawing.

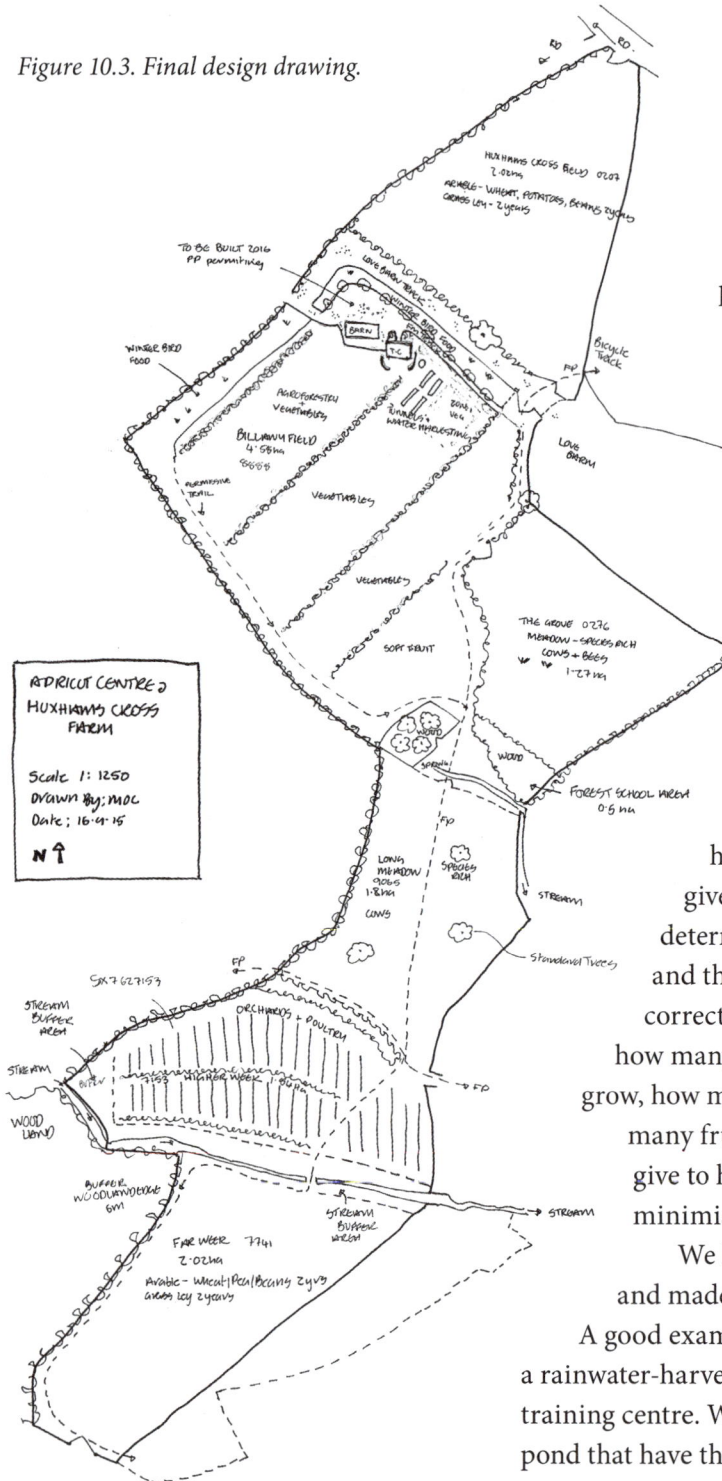

are therefore right where they should be on a biodynamic farm – in the centre of it.

Once we had placed the elements on our giant floor map, we spent some time imagining the flow of work and people on the emerging new farm. We had prepared lots of figures beforehand relating to the inputs required. For instance our two-hectare field that was planned for small-scale grain production would produce, we estimated, approximately five tonnes of YQ population wheat per year and would feed approximately 100 chickens with some extra protein input. Five tonnes of wheat require $10m^3$ of barn storage space. The research we had done in the input–output analysis described above helped us decide how much land to give to each element in the farm, and to determine the carrying capacity of the land, and therefore helped us scale everything correctly. This guided our decisions as to how many chickens to have, how much grain to grow, how many cows the farm could support, how many fruit trees to plant and how much area to give to horticultural crops – all with the aim of minimising our bought-in inputs.

We looked at the outputs from the system and made sure they would be used on the site. A good example of this was the installation of a rainwater-harvesting system next to the barn and training centre. We planned to install a tank and a pond that have the capacity to hold 125,000 litres of water. These would collect water from the roof of the two buildings, which would then be fed through irrigation pipes to the polytunnels and vegetable beds. We thought that if the stored rainwater ran short (as it has in two out of the three summers we have been on the farm so far), then we could top up the tank and pond from the springs on the farm, or as a last resort from mains water. This arrangement would give resilience to the water system. We planned to slow down the rate of runoff of the rain falling on to the vegetable field through the inclusion of agroforestry rows running across the contours, which would enable the water to penetrate down through the subsoil to replenish the aquifers and the springs. These rows of trees would also slow down the wind, thereby slowing down the loss of water by evapotranspiration from the plants and evaporation from the soil. We planned to build up the organic matter content of the soil over the years with the use of green manures; this would also increase the capacity of the soil to hold on to water and reduce the need for irrigation in the longer term. Elements of the model were shuffled around until it worked and flowed.

We repeated the whole process a few weeks later, drilling down into the details. For instance, where we had just written 'orchard' on the original design we now planned out the direction of the rows, the sequential cropping of fruit, the positioning of windbreaks. This work in turn led to the creation of a detailed business plan and implementation timetable outside the workshop.

Working as a group brought unexpected and wonderful creativity, pushing the boundaries of us as a team, and making the design richer and more playful. Working as a team also brought up conflict-rich hotspots. Particular hotspots concerned the need for biodiversity on the new farm versus the need to grow food. This conflict was a microcosm of the conflict that arises in relation to biodiversity around the world in general, so the way that Huxhams Cross Farm planned to address this issue is and was key. Once carefully unpacked, such conflicts can lead to rich and creative solutions to the need to grow food but not at the price of decimating the wildlife. At Huxhams Cross Farm we left one field for a year of observation, carefully mapping out the beautiful orchids and other wild flowers on the site.

We subsequently planted extensive orchards where these flowers were not present and managed the rest as a wild flower meadow with conservation grazing with Shetland cows. The cows have also been fenced out of the area where the spring water rises, in order to cut down the contamination of one of the headwaters of the river Dart. The permaculture principle of 'small and slow solutions' was especially useful for these hotspots of problem-solving. Figure 10.3 shows the final design.

Implementation

The implementation of the design began in the autumn of 2015. We 'sculpted' the design on paper into the actual landscape of Huxhams Cross Farm and tweaked it as the farm developed. The very first thing we did, even before we had signed a lease on the farm, was to plough and harrow the arable fields and put them down to a very rich mixture of deep-

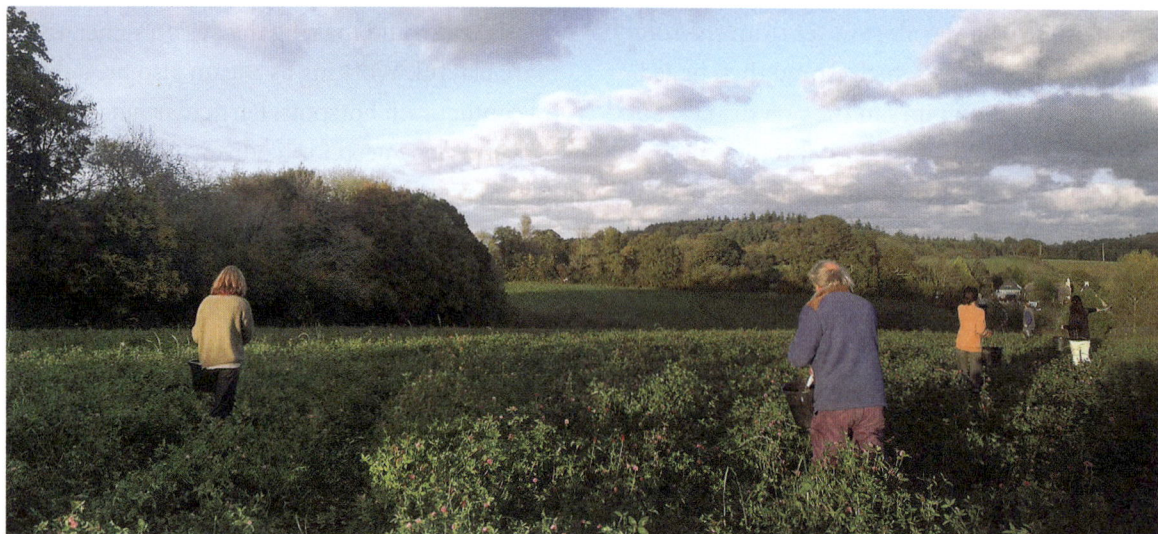

Figure 10.4: Putting on the biodynamic preparations.

rooting green manures to start to recondition the soil and bring it back to life. The green manures were designed specifically for the soils at Huxhams Cross Farm.

In the first winter we planted 2,000 trees, mostly on the contour in agroforestry rows; we planted a further 2,000 soft fruit plants and fruit trees and built the barn. In year two we contracted the use of a keyline plough to ease compaction in some of the fields, put a flock of hens on the farm to bring in some much needed cash and hired a contractor to put in our first population YQ wheat crop and put up the barn. In year three we planted our first vegetable crops in between the agroforestry rows, picked the first crops of strawberries and harvested our first YQ wheat crop. We also installed the rainwater-harvesting systems, including the tank and pond that enabled us to store 125,000 litres of water at any given time – or would have done if it had not stopped raining the moment we installed them! We had to wait until the following year to

have collected enough water for our polytunnels. We put up six second-hand polytunnels, which gave us 600m² of covered growing space. Our training centre was also built in 2018, allowing us to start to deliver our well-being programme of therapy for children on the farm. We brought the farm into full production in the fourth year after signing the lease, although the fruit crops will take five to seven years to come into full cropping.

The toolkit we used was the permaculture design methodology. What we have created is a biodynamic farm incorporating agroforestry and regenerative agriculture and welcoming people back to the land.

The biodynamic conversion process

Bringing the soil back to life: the day we signed the lease for the farm tenancy we applied for the farm to go into the biodynamic conversion process. We compiled a 'conversion plan' for the farm explaining how we were planning to convert it to a biodynamic system over the next three years. How were we going to bring the soil back to life? The first stage was putting the arable fields down to deep-rooting green manures and applying the horn manure preparation to reintroduce microorganisms into the soil (Figure 10.4); it was this that brought the first group of 25 people to the fields – probably more people than these fields had seen in a long time.

Bringing people back to the farm in this way, to quote Steiner, is 'giving the farm its soul'. A good place to begin the new farm journey.

In the summer months we put on the horn silica preparation and made our first compost heap and put the compost preparations into it. The second autumn, we started making the biodynamic preparations on the farm, the local biodynamic group leading the way. We filled cow horns and buried them and made some of the compost preparations. We continued to put the horn manure on once or twice per autumn and recently we have added to that the 'cowpat

Figure 10.5: *The author at Huxhams Cross Farm.*

pit' preparation. The latter contains all of the compost preparations, so by adding it to the horn manure mix we can apply the compost preparations to all of the arable fields, since we haven't had enough compost to put on them. We use the biodynamic calendar to aid in the choice of day to apply the preparations, aiming for a root day for the horn manure. We apply horn silica to the crops in the summer months. We have just invested in a flow form to stir larger amounts of preparations and apply them with a sprayer on the back of our small tractor.

Our food does have a distinctive taste, or 'terroir'. It is highly regarded by our customers. It is delicious and has great keeping qualities. Our soil went from being the worst soil on the Dartington Hall Estate to the best-performing soil in three short years.

Creating the farm organism is a slow process. We have been around the full rotation only once in five years.

Designing and implementing the agroforestry systems

Our main vegetable cropping field is called Billany and is four hectares. It has the best soil on the farm and faces south. The plan was and is to grow vegetables with polytunnels and some soft fruit in this part of the farm. It is lovely and sunny but exposed to the wind. The soil is clay over shillet, basically a free-draining clay, but most of the top soil is at the bottom of the hill after years of soil erosion.

We chose hazel coppice as our agroforestry tree species because of the exposure and our presumption that we would need wood to heat our training centre. As it turned out, the

Figure 10.6: Drone picture of Huxhams Cross Farm 2016.

new training centre was built with such good insulation that a wood-burning stove would have made it too hot. Instead we will use the coppice to make ramial wood chips to improve the soil carbon content. Our next decision was which way to plant the alley rows: on a north–south axis up and down the slope or on the east–west axis on the contour? This was an agonising decision, since each solution had its merits, but we knew that once we had planted the trees it would be difficult to change their arrangement.

We used a permaculture design tool called 'Strengths, weakness, opportunities and challenges' (SWOC) to help us make our decision. Because we had done so much work

to clarify the functions, aims and objectives of the farm, the choice was quite simple in the end. We decided to plant the trees on the contour running across the field with 28m spacing between them. Planting east to west means that we do give some shade to our crops, but because the farm is on a slope this is minimal. In order not to shade our crops out we kept the spacing very wide at 28m, so we could at a later date add another row in the middle, which would make the rows 14m apart. The fact the trees run across the slope slows the movement of water and topsoil down the slope, helping water to percolate down into the subsoil and aquifer below. The trees break the prevailing wind and this in turn reduces evapotranspiration from the crops in hot windy weather. The agroforestry rows create a human-scale farm in which to work; we can see right into all of the cropping areas. The agroforestry alleyways are three metres wide, are full of long grass and are a perfect home for our functional biodiversity predators as well as many linnets. They bring a lot of 'edge' into the cropping areas. The way they break up the space makes it easier to plan our rotations in the alleyways; we made them the right size to fit the standard sizes of crop covers and the length of pipe our irrigation would run.

Because the agroforestry rows have been planted on the contour, the tractor work is carried out on the camber – on a slight slope. Most tractor implements work better on the flat, so most farmers carry out cultivations up and down a slope, even though this causes problems with soil erosion. We have found that all but one of our cultivations can be done adequately across the slope; the exception is mechanical weeding. To compensate for the latter we have invested in pedestrian wheel hoes that are fast and efficient to use by hand. Had the slope been steeper, we would have made a different decision. After four years of cultivation two of our agroforestry alleyways have become home to no dig intensive beds that run up and down the slope (see Figures 10.6 and 10.7).

In our meadows we planted standard perry pear trees to add a potential crop in 20 years' time and to add more biodiversity and bee fodder to our wetland meadow. In the fullness of time they will offer shade to our two cows grazing in this meadow.

Figure 10.7: Drone picture of Huxhams Cross Farm 2021.

Photo: Christian Kay

We are partners in Broadlears Agroforestry Field on the Dartington Hall Estate. This is a 20-hectare field with agroforestry rows 20m apart. We have planted a third of these rows with 800 fruit trees. Luscombe have planted elderflower on a third of the rows for their elderflower cordial, and the London Peppercorn Company have planted a third with Sichuan peppers. The arable cropping space in between is farmed by Jon and Lynne Perkin of Old Parsonage Farm; they grow wheat, hemp and oats there. The wheat is the genetically diverse wheat produced for Dartington Mill: the YQ population wheat, Cornovii, a new wheat created by Fred Price of Gothelney Farm in Somerset. Some of the oats are sold to the Lush cosmetics company as fresh oats for their hand cream. The field is registered organic. All the partners pay rent for their allocated areas and have signed a lease that reflects the shared values underpinning the way the crops are managed in the field.

Dartington Mill

A few years after we took on the lease for Huxhams Cross Farm it became clear that it was not economically viable to grow two hectares of YQ wheat. At the same time, Old Parsonage Farm at Dartington Hall and the Almond Thief bakery, close by, were keen to add value to their grain and to access local flour for baking. Together the three companies formed Dartington Mill, out of the Grown in Totnes not-for-profit organisation. Old Parsonage Farm has 120 hectares of arable land (out of 180 hectares in total), on which the Perkins grow a wide range of grains. Dan Mifsud at the Almond Thief bakes artisan sourdough loaves and shares a building with the New Lion Brewery. Together we bought a new mill from the US which mills the grain slowly without overheating it, preserving its taste and nutritional value. The Apricot Centre sells its grain to the mill; the mill processes it and then sells it on to the bakery, occasionally to the brewery, or the Apricot Centre buys it back. The Apricot Centre puts the flour in bags and retails it. Dartington Mill trades under the name of 'Reclaim the Grain', since our aim is to shorten the supply chain and decommodify and relocalise our grain production.

What has the farm achieved after five years?

After five years of hard work to bring the farm into being we have achieved a great deal. We are working towards a closed loop farming system, growing many of the resources we need on the farm and putting back any waste products in the form of compost. We are delivering 500 hours per year of individual therapy to children and run an after school farm club and school visits. We welcome approximately 1,000 people to the farm each year on visits and tours. We train approximately 40–50 people per year in permaculture, biodynamic farming and growing and agroforestry. We are about to scale up our apprenticeship scheme so we can train 20–30 people per year. They will be placed in farms across Devon and trained at levels three and four in regenerative food systems, encompassing the systems outlined in this book.

We are financially self-sufficient and employ six people on the farm, have three apprentices and employ five full-time-equivalent in the well-being service and one person in business

Farming as the key to net zero carbon emissions by Philip Franses

Philip Franses works for Flow Partnership, an NGO that partners with other organisations around the world to reinstate and restore watersheds and the small water cycle.

Many people have the idea that we somehow passively own carbon in a reserve of fossil fuels and have expended this resource in the burning of coal and oil. For instance, common arguments around the ambition to reach net zero carbon emissions by 2030 often refer to the heating of buildings, changing to natural energy sources, and methods of transport – without even mentioning the land. How we farm the land will be a crucial element in the future management of carbon emissions and hence the response to climate change.

Carbon, it is believed, is made in the hot centres of stars, through a chain of unlikely reactions known as the triple alpha process. In 1953 the renowned astronomer Sir Fred Hoyle predicted a then unknown excited state of carbon must act as a stepping stone for the production of stable carbon. This required a number of different chemical constants to be exactly aligned. Were this not so, then the earth would have no carbon to provide the basis for organic life. Carbon naturally cycles between organic life, sediments, soils, the atmosphere and the ocean; photosynthesis and respiration are part of this cycle.

We participate in the carbon cycle and without it we would not exist. The climate is changing because this cycle has become impoverished. When we restore the cycling of carbon in the way we manage a farm with healthy, porous, nutrient-rich soils full of organic life, the carbon picture changes dramatically.

Impact assessments carried out by the Apricot Centre team after five years of farming at Huxhams Cross Farm, using the farm carbon toolkit, have measured that 63 tonnes of carbon per year, over and above what the farm has used, are sequestered from the atmosphere into the farm's 13 hectares of soil. To put this in perspective, agricultural land is by far the biggest sector of the earth's land surface, covering around 5 billion hectares. The rate of carbon sequestration achieved by this farm would translate worldwide into a cycling of 25 billion tonnes of carbon per year.

In burning fossil fuels to fuel industry and changing land use (deforestation etc.), we are emitting about 40 billion tonnes of carbon into the atmosphere per year. So the lack of capacity to hold carbon in the land in the way we farm is just over 50 per cent of the total net emissions of carbon into the atmosphere which are driving global warming.

Such results are similar to findings about potential carbon sequestration in wetlands, another key player in a healthy carbon cycle. Natural England estimates carbon sequestration in wetlands as amounting to up to 14.5 tonnes per hectare per year. By impairing the carbon cycle through removing wetlands and through monocrop farming we are inadvertently taking carbon away from its natural use in the soil food web.

The IPCC goals are to reduce carbon emissions by 2030 to 25–30 billion tonnes and to reach net zero emissions by 2050. Regenerative agriculture over large swathes of land could play a huge role in restoring the natural carbon cycle so that it will have the capacity to sequester these amounts of carbon. It was a welcome surprise to me that innovative farming practices offer a proven method of carbon sequestration.[8]

activities such as accounts and administration. That is twelve full-time-equivalent people. The farm also looks very beautiful and the produce tastes divine. So how has our farm in its short life met the four challenges it was designed to meet?

Climate change mitigation

We have planted 2,000 hazel trees and 1,300 fruit trees. We have converted the land to a fully biodynamic farming system and piled organic matter into the soil via green manures. We have installed a five kW array of PV panels on the roof of the barn. All of our deliveries are carried out within a 30 km radius and we have reduced our plastic use to reusable bags only and we use them for salads and pre-packed greens only. We do buy in extra supplies, but our policy is that we source locally wherever possible from within the UK and then from Europe if necessary, but never shipped by air.

We have worked with the farm carbon toolkit[6] and found that we sequester 62 tonnes of carbon dioxide more per year than the farm puts out. That is, the farm sequesters twice as much carbon as it uses, or almost five tonnes per hectare per year. To put this in context, the average person in the UK expends 5.3 tonnes of carbon dioxide per year. At Huxhams Cross Farm we are still paying the carbon debt for our barns and training rooms. Use of the farm carbon toolkit has highlighted where we can reduce our carbon the most (see Figure 10.8), and that is in the organic chicken feed that we buy in – we still need to supplement 50 per cent of our chicken feed with bought-in layers pellets – and in the fuel for our delivery vehicle. Our next step will be to switch to an electric delivery vehicle and

to source locally produced high-protein chicken feed. What have been most effective in terms of carbon sequestration are our deep-rooting green manures (see Figure 10.9).

One research project carried out on the estate has highlighted how Huxhams Cross Farm's soil has formed the highest amount of soil aggregates among all the farms on the Dartington Hall Estate. These are what sequester carbon on a stable long-term basis; biodynamic farms are particularly good at forming them, as demonstrated by research at FiBL.[7] These soil aggregates are formed by the relatively high levels of bacteria and fungi in the soil.

Climate change adaptation

Since we took on the farm in 2015, the weather events the farm team have had to deal with have been as follows:

- 2015: warmest year on record
- 2016: monthly extremes
- 2017: fifth warmest year on record
- 2018: the 'Beast from the East' followed by summer drought
- 2019: warmer, wetter and sunnier than average; hottest day ever recorded in the UK in July
- 2020: Storms Ciara and Dennis, producing heavy rainfall and flooding; driest May on record
- 2021: frostiest April on record; wettest and coldest May on record.

At the start of our cropping season in 2017 we had such extreme weather that I found myself ranting that we were not yet ready for climate change. However, the farm has become resilient

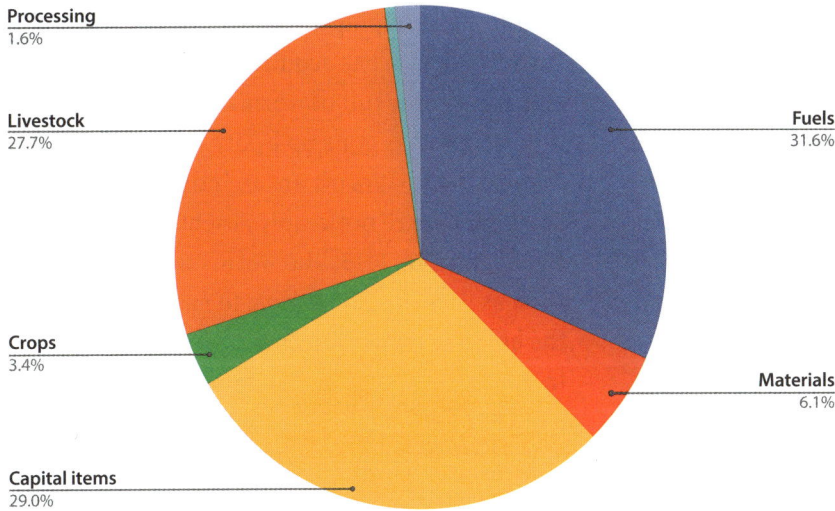

Processing
1.6%

Livestock
27.7%

Crops
3.4%

Capital items
29.0%

Fuels
31.6%

Materials
6.1%

Figure 10.8.
Pie chart of carbon use.
CO_2e emissions from the
farm. Fuels, livestock
and capital items are the
biggest contributors of
GHG emissions.

Figure 10.9.
Pie chart of carbon
sequestration.
Total CO_2e sequestration
on farm.

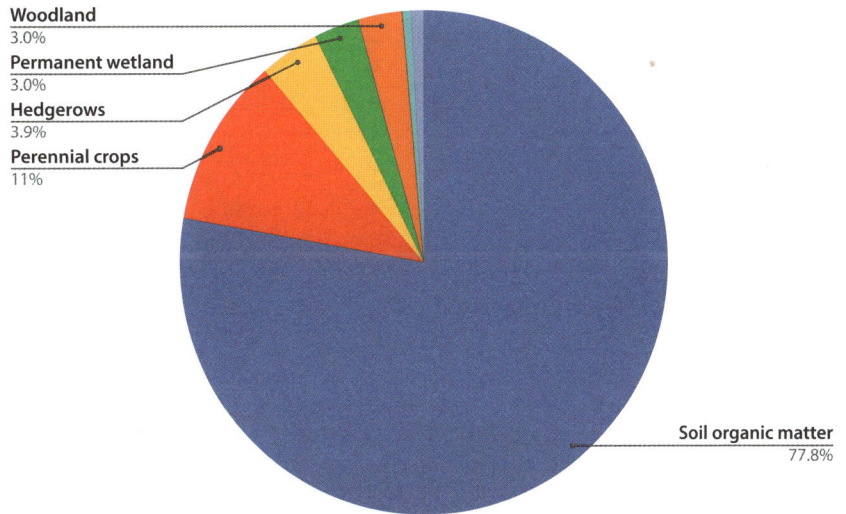

Woodland
3.0%

Permanent wetland
3.0%

Hedgerows
3.9%

Perennial crops
11%

Soil organic matter
77.8%

to such weather events and we have managed to cope. We have managed to continue producing high quantities of good-quality fruit, vegetables and eggs. In 2020, when wheat yields in the UK went down by 30 per cent, our grain yield was the same as in the previous years. We installed the tank and ponds with capacity to store 125,000 litres of rainwater; in the first few years the water ran out every summer and we had to resort to mains water for a few weeks. By the summer of 2020 we were managing totally on the stored water. The amount of mains water we use each year will vary according to the year's rainfall distribution. Our soil has increased its organic matter by a huge 25 per cent and is covered most of the year with mulch and green manures, which makes it more resilient to heavy rainfall.

Offsetting biodiversity loss

With careful grazing and hedgerow management, we have restored three hectares of wetland meadows that had been neglected for the last 30 years and that in most farms would have been drained. The orchid count has gone up by a factor of eight, as have the insect, worm and bird populations. When you walk through the meadows in the evenings they are alive with bats and owls, and the summer of 2020 saw an explosion of crickets and meadow brown butterflies.

On the cropping areas of the farm, as the soil has recovered it has acquired a better structure and is also now full of worms. The worm count on the farm has gone up by 50–400 per cent in most fields. Worms are an indicator of soil health and were noticeably absent when we took on the farm in 2015. The worms turned up as soon as they had something to eat; they are near the bottom of the food chain, so they themselves in turn become food. The agroforestry rows are now full of linnets as well as the voles and mice that the cats bring in.

Figure 10.10: Huxhams Cross Farm wellbeing pod and intensive beds.

The RSPB did bird surveys in 2015, when 19 species were noted, and in 2020, when 28 species were noted – approximately a 30 per cent increase. Our green manures of clover and buckwheat are alive with bees when they are in flower. Predatory insects turn up to feed on our pests with welcome regularity.

Producing enough food for everyone

The farm is productive: at Huxhams Cross Farm we are producing enough food for approximately 250 families per week in terms of vegetables, fruit and eggs, plus approximately five to six tonnes of wheat per year (one tonne per acre), some of which goes to feed the chickens and most of which is milled into flour and sold either to a bakery or direct to customers.

The yields and economic performance of the farm in 2020 were as follows. We harvested a total of 15.4 tonnes of fruit and veg – almost a 20 per cent increase from 2019. In addition, we collected a total of 38,500 eggs (or 6,410 boxes), 2.4 tonnes of hay (used as animal feed over winter), six tonnes of wheat and 2.3 tonnes of

straw. The fruit and veg are mostly sold directly to the consumer, via our online shop and weekly market. Less than 10 per cent was sold wholesale to local restaurants and small shops in 2020. Surplus produce was processed and sold as jams, chutneys and juices, or donated to the local food bank. Our local food bank is called Food in the Community and provides fresh food to those in need. A small amount of waste was composted on site or fed to the chickens.

We grew a range of 100 different crops and varieties; 82 vegetable crops and varieties and 18 different fruits. We made about 30 different products using our preserving equipment and processed the wheat to sell as flour (wholemeal and white), flaked wheat and wheat berries (whole grain used as a rice substitute). Of the waste products from Dartington Mill, the bran is sold to local mushroom producers, who use it as a substrate on which to grow mushrooms, and anything left is used as chicken feed. Straw and hay were used as animal feed, mulch and animal bedding, and some was sold to customers. We

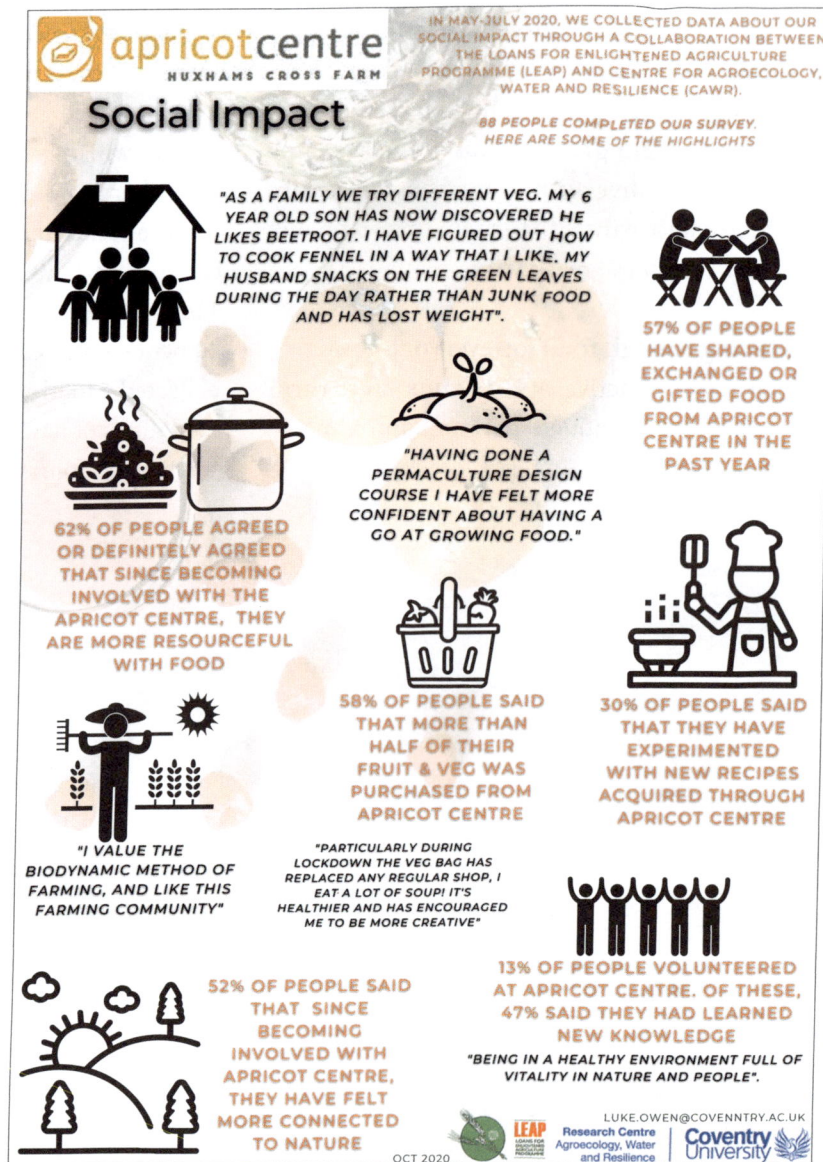

Figure 10.11: Infographic from CAWR at Coventry University.

have a stable flock of approximately 150 White Leghorn chickens providing fertility, pest control and a total of £16,000 of income per annum.

All our produce is sold within a 30 km radius of the farm. The majority is sold directly to the consumer. In 2020, approximately 75 per cent of sales occurred via the online shop, 18 per cent via the local market and seven per cent was sold wholesale. Owing to COVID-19, there was a 350 per cent increase in sales on the online shop. The local market was closed for two months. However, despite two months of no sales there, total market sales for the year increased by four per cent compared with 2019. When the market reopened (June 2020), our sales more than doubled.

By autumn 2020 our sales had stabilised, with less variation from month to month.

The total income from the online shop and market sales had increased from £6,500 per month in 2019 to £17,500 per month in 2020. This is a 260 per cent increase. The number of customers also doubled from 2019 on both the online shop and the market.

We also buy organic produce to supplement our own. This is as local as possible, but we do import from further afield, mainly for fruit in the winter months, including from Spain, Italy and the Dominican Republic (for bananas). In 2020, bought-in produce accounted for a third of our total sales.

The value of our crops increased in 2020 because we increased our salad and herb production, these being high-value crops. The average value of produce increased from £4.47/kg to £7.85/kg.

We have carried out a social impact study of our food in partnership with the Centre of Agroecology (CAWR) at Coventry University (Figure 10.11). They found out that our customers wasted less food (63 per cent), felt more connected to the origin of their food (72 per cent) and ate more than the national daily average of fresh fruit and vegetables (91 per cent). One lovely quote: 'As a family we try more veg. My six-year-old son now finds that he likes beetroot. I have figured out how to cook fennel so I like it. My husband now snacks on the green leaves rather than junk food and has lost weight.' ■

1 Attributed to Mark Twain.

2 Quotation from D. Holmgren, *Permaculture: Principles and Pathways Beyond Sustainability*, Permanent, East Meon, UK, 2002.

3 'Devon banks' are a particular form of ancient hedgerow found in Devon. The soil is mounded up to a height of one metre and lined either side with stone facing, using stones from the fields. The tops of the banks are planted with trees that are regularly coppiced. These ancient hedgerows are full of biodiversity and up to 800 years old.

4 M. Large and S. Briault (eds), *Free, Equal and Mutual: Rebalancing Society for the Common Good*, Hawthorn Press, Stroud, 2018.

5 https://www.biodynamiclandtrust.org.uk

6 https://www.farmcarbontoolkit.org.uk/

7 P. Mader, A. Fliessbach, D. Dubois, L. Gunst, P. Fried and U. Niggli, 'Soil Fertility and Biodiversity in Organic Farming', *Science* 296, 2002, pp. 1694–1697.

8 C. Harvey and N. Gronewold, 'CO$_2$ Emissions Will Break Another Record in 2019', *Scientific American*, 4 December 2019; Natural England, 'Carbon Storage by Habitat: Review of the Evidence of the Impacts of Management Decisions and Condition of Carbon Stores and Sources', NERR043, 2012; J. Rogelj et al., 'Mitigation Pathways Compatible with 1.5°C in the Context of Sustainable Development', in *Global Warming of 1.5°C*, ed. V. Masson-Delmotte et al., World Meteorological Organization, Geneva, 2018.

Chapter 11

Fermenting the revolution

Hey farmer farmer put away the DDT now,

give me spots on my apples

but leave me the birds and the bees, please

don't it always seem to go

that we don't know what we've got till it's gone

they paved paradise and put up a parking lot

Joni Mitchell – 'Big Yellow Taxi'

How might a sustainable agricultural revolution be brought about?

A huge and complex set of challenges face food production in the next 30 years, and these require equally huge and complex responses. These will be multiple and diverse responses, because sustainability is a process rather than a state and we are in a process of transitioning towards a more sustainable food and farming system. It might be a 1,000-hectare monocrop arable farm on the East Anglian plain moving to a no-tillage system and cutting down on nitrate fertilisers, or an organic farmer in agroforestry alleyways to improve yields, microclimates and wildlife habitats. It might be an Incredible Edible local group planting up the beds in front of a public library with tomatoes instead of marigolds,

or it might be someone in a city planting a container of salad on their balcony and shopping in the local farmers' market. These are all steps towards resilient and sustainable food systems.

When complex systems change they go through a process of 'chaos', and multiple changes occur at the same time. Chaos theory tells us that complex systems have 'emergent properties' as they transition from one thing to another. The current transition from industrial to sustainable food systems can be viewed in this light. Charles Massy calls the mindset of the sustainable farmer an 'emergent mind'.[1]

The IPCC's stated aim is zero emissions by 2050 to achieve a 50 per cent chance of avoiding runaway climate change. Net zero emissions by 2030, as more radical movements demand, would require faster systemic change. The changes will need to meet all four challenges – mitigation of climate change, adaptation to climate change, offsetting biodiversity and producing healthy food – at the same time, since we will not have time to address them one problem at a time.

Part of the solution is to use the toolkit of sustainable methodologies described in this book and begin to farm, garden and use our public spaces differently. Different methods will resonate with different people, and different combinations will be used. Some steps will be small and some may be large. But all steps along the path towards producing food in a more sustainable way are important and should be celebrated.

It may seem improbable that the majority of food producers in the world could ever be sustainable rather than industrial. We could ask ourselves, 'What if the majority of food producers moved towards sustainable food systems? What might that look like and how might it come about?' Improbable though the prospect seems, the FAO suggests that a transition of the majority of farms to…

…agroecology systems… should start now, because farms and food production are in the unique position of being able to mitigate climate change, by reducing its carbon emissions and sequestering carbon in the soil and into trees. At the same time farms and food production needs to increase by 60 per cent whilst adapting to climate change if the world is to avoid food insecurity.[2]

How might such a huge change come about?

Our own food choices

Everybody has to eat. We can change our shopping habits and buy our food from sustainable local sources. We can learn to cook from scratch, grow some of our own food, buy shares in a land trust and volunteer or work on a farm. We can join the local branch of La Via Campesina and support them in lobbying governments for sustainable change. We can join regional biodynamic, organic or permaculture groups. Vandana Shiva said, 'the most revolutionary act any of us can do right now is to buy, grow and eat local food'.[3]

Eating seasonally reconnects us, as consumers, with nature and the turning of the seasons and we know that this natural connection is good for us. At the time of writing, in the midst of the COVID-19 epidemic, there has been a surge of interest in the importance of nature connection, of walking and experiencing nature during lockdown. Eating healthy seasonal food has also had a surge in demand during this period. Wendell Berry expressed this kind of connection beautifully when he wrote:

Eating with the fullest pleasure – pleasure, that is, that does not depend upon ignorance – is perhaps the profoundest enactment of our connection with the world. In this pleasure we experience and celebrate our dependence and our gratitude, for we are living from mystery, from creatures we did not make and powers we cannot comprehend.[4]

Locally sourced sustainable food is expensive; in the global north it costs more than industrially produced food and is perceived as middle class. It's assumed that people on low incomes will not be able to afford to eat this kind of food. However, if a switch is made to a plant-based diet, if meals are cooked from scratch, if less processed foods are eaten and if food cooperatives are set up to source food in bulk, then healthy, local sustainable food can be affordable.

Eve Balfour described organic food as a national health service. If the health dividends were factored in, it might be more cost effective to use nutrient-dense food as a form of preventative medicine. There are huge structural issues in play – from the 'food deserts' in low-income areas, to poor cooking facilities and dining space in modern homes, and lack of life skills in preparing food from scratch – which cannot be addressed on a purely individual level.

Food retailers and caterers' choices

Retailers, caterers, chefs and cooks can source sustainable food. This is easier said than done. Local sustainable food is more expensive, it takes more preparation and it often arrives muddy and needing scrubbing and peeling, unlike the washed, chopped and vacuum-packed vegetables that arrive from the industrial food sector. The supply is seasonal and therefore erratic: a chef will not be able to put cucumber on the menu in the UK in January. However, with retraining, chefs and cooks can learn to adapt recipes and menus in order to use seasonal food.

Rob Hopkins describes a visit to a school in Mouans-Sartoux in France in 2019. In 2012, this town of 10,000 people took on a seven-hectare site where they now grow organic vegetables and fruit. They supply all the local school canteens, government offices and day-care centres with fresh organic fruit and vegetables. This helps the children develop healthy eating habits and helps them concentrate better. The children go to the market garden to participate in growing the food and learning about healthy food and the environment. The integration of the market garden into the school system has given a boost to the local economy, changed many families' eating habits and deepened understanding of climate change. An organisation was set up, the Sustainable Food Education House (Maison d'Éducation à l'Alimentation Durable – MEAD), with a seminar and visitor centre to show others how to train communities to create market gardens and use these to transform school food and educate children. They help other schools and public institutions to make changes in their kitchens and to train staff in different approaches to cooking and menu planning.[5]

In France the government has instituted a policy that half of the procurement of food in the public sector – schools and hospitals, prisons and government buildings – will be organic or locally sourced by 2022. This will boost the French agricultural sector and improve diets where it is most needed.

Food producers' choices

Farmers, growers, allotment-holders, smallholders and community gardens can make the decision to transition to sustainable and regenerative systems. Once the decision has been made, the transition needs to be carefully planned and support, advice and training to be sought. Often in the first two years of transition, as the farm and land adjusts, there can be some difficult moments and crop failures before the ecology systems start to recover and to fully function. Regenerative farming systems are usually more economically viable, are more complex, but bring more skill and interest to farming and growing. The farmers and growers become more socially connected with their customers and colleagues, and biodiversity returns quickly.

Changing our minds

In all of our farming systems the mindset of the farmer makes a big difference to the outcomes on the farm. Rabindranath Tagore, the poet, suggests a that frame of mind 'guides our attempts to establish relations with the universe either by conquest or by union, either through the cultivation of power or through that of sympathy'.[6]

The farmers who have made the transition to sustainable farming systems share with indigenous farming peoples a similar ethos of the psychology of abundance, or, as Tagore puts it, of 'union and sympathy with the natural world' rather than seeking to control it. The attitude of these farmers is to have a strong connection and working relationship with nature. They create circular, reciprocal systems that give something back and replenish soils. They create positive feedback loops. They don't need to control the environment around them and are curious to observe and learn from their systems and willing to make the odd mistake.

Sustainable farmers tend to be empathic,

love diversity and are generous. They know that nature has the resources both to provide for their customers and to sustain biodiversity on their farms. The psychology of scarcity, on the other hand, is full of the fear that there will not be enough food. The linear input–output model of industrial farming puts the emphasis on only one output of the system, that of yield and profit, and ignores the cost of clearing up the damage that is done. Industrial farmers outsource to others and to nature itself the burden of dealing with the damage they do to the environment.

Massy describes this as a paradigm change: 'a switch from the mechanical mind to the emergent mind involves dismantling and then rebuilding the entire superstructure of one's belief system and worldview. [This] triggers a strong ethical and moral element and often an openness to the spiritual element.'[7]

I have trained people in the sustainable farming techniques described in this book over many years. I usually start each course with the question 'Have you had a strong experience in nature in your life?' The answer is invariably 'yes'. The paradigm shift is often brought about by a strong experience in nature that changes one's perception of the world.

In Part 1 we looked at the ecopsychologists' approach to understanding the psychology of abundance versus the psychology of scarcity and why someone occupies one mindset or the other. Ecopsychologists' conjecture is that it may be the trauma of past food shortages that keeps us so addicted to industrial food systems, that is, to having lots of cheap food available all the time, and to inhabiting the psychology of scarcity rather than the psychology of abundance. The trauma therapist Dr Gabor Mate in his film *The Wisdom of Trauma* says,

We are going as far as destroying the earth because of our addictions… the disconnect from the earth has to do with the disconnect from our own bodies the two are together… we call our earth 'mother earth', but look what we are doing to her. It's like mother hatred almost. It speaks to a societal blindness and passivity that is itself a mark of collective trauma.[8]

Rust suggests that in order to heal an unhealthy relationship with food an individual must first process the trauma they have experienced and reconnect the sensations in their body, such as hunger, or feeling full after a meal, to their eating habits. That means developing an awareness of when they are hungry and when they are full. What might this kind of process look like on a collective level?[9]

Can any parallels be drawn between an individual healing from trauma, through giving time to processing the trauma and reconnecting to their body to overcome addiction, and a collective trauma and a similar recovery on a collective scale? Are farmers learning to reconnect to the soil and the land they live on, healing it and regenerating the five landscape functions, a little like an individual might reconnect to their body, creating a healthy relationship with it? Can we perceive sustainable farmers as land healers or therapists? Are they healing their own trauma or are they collective therapists? Are they helping us, the consumers of food, to reconnect to nature via food, delivered as fresh, healthy, seasonal food? Are they healing our damaged

food systems by means of their farming skills and their unique attitude towards nature?

These farmers are highly skilled in their farming techniques and have mastered their art. They are creating new complex systems unique to their own time and place, using the techniques and methods described in this book.

Policy-level choices

On a government level there are policy changes that can be made. Switzerland and Cuba both have large amounts of land under agroecological systems of farming. What can be learned from these examples about how government policy can support the transition?

In Switzerland it has been government policy to subsidise and support farms to become more sustainable; the government values the benefit to the country's environment and also the capacity to be self-reliant in times of crisis. The FAO proposes a realignment of policy to reward sustainable agriculture and provide financial support and training for those moving towards sustainable agriculture systems.

Cuba made a rapid change to sustainable farming methods because of the crisis in their oil supply after the fall of the Berlin Wall and subsequent collapse of the USSR. Cuba was left isolated and unable to trade for oil with any other states, so it made a rapid shift to sustainable low-carbon forms of food production. This was very difficult for the interim three years when people went hungry, but once the transition was made, the general health of the population improved.[10] Food production became more urbanised and was relocalised. Sugar and tobacco, the main export crops, were replaced with healthy fresh food for local people.

Political policy is key to the expansion of sustainable farming – unless we want to wait for a crisis in either oil supplies or climate change to force sudden change. Policy levers to change farming practice come in the form of subsidies for good practice, the 'polluter pays' concept, setting targets and providing free advice and support.

The use of subsidies to influence change may be illustrated by the new Agricultural Bill passed in the UK (2020) upon exiting the EU. This bill and the new policy were developed with many stakeholders, including the Landworkers' Alliance (the UK member of La Via Campesina). The UK had since 1973 subscribed to the Common Agricultural Policy (CAP), which paid area-based subsidies to all farmers in the UK. In future, in the UK, farm subsidies will only be paid for 'public goods' provided by farmers. 'Public goods' are defined as: supporting biodiversity, carbon sequestration, clean water and clean air, reducing environmental pollution, and public engagement. This scheme is called ELMS, the Environment Land Management Scheme.

The 'polluter pays' principle has been used in many countries.The plastic bag charge imposed at many supermarkets has radically changed shopping behaviour. This approach could be applied to farms: those using nitrate fertilisers and pesticides could pay a tax that would pay for cleaning these substances out of waterways and for biodiversity loss. Jules Pretty has calculated that industrial farms cost £209 per hectare per year to clean up. In the UK a climate change levy is charged for carbon emissions from industry, but horticulture and agriculture are exempt. One idea often cited is that industrial farmers should have to list any 'P' numbers (pesticide numbers) on their produce, be inspected annually to ensure they are complying with the legislation and undergo audit trails to prove what they have used and when. The requirement of a licence to use pesticides would discourage their use. A requirement for all farmers to undertake an audit trail for the use of pesticides and to be annually inspected may sound unlikely, and yet all registered organic and biodynamic farmers have, each year, to pay for an inspection and to carry out audit trails for all their bought-in seeds, compost and products in order to prove that they are organic or biodynamic and have the right to sell their food under the corresponding symbol.

Target-setting is another policy lever. In the UK the government has set a target of planting 30,000 hectares of trees per year, starting in 2024, to increase the woodland cover from 13 per cent to 17 per cent. Currently, 70 per cent of land in the UK is farmland, so this target could also include the planting of agroforestry systems on farms. Planting trees in the form of agroforestry would increase food production rather than taking land out of food production to become forest.

Governments can provide free training and advisory support for farmers converting to sustainable systems. Such training and advice were provided after World War II to bring about large-scale change to industrial farming. There is a chronic lack of skills in sustainable food and farming systems, so this kind of support could be very effective.

At the international level the UN has clearly stated that its aim is that all farming should shift towards the agroecology model because this both mitigates and adapts to climate change while addressing food shortages. The 17 Sustainable Development Goals (SDGs) were set and agreed by the UN in 2015 as a blueprint of goals to be achieved by 2030. They include goals such as eradicating hunger, addressing climate change and biodiversity loss, reducing plastic waste in the oceans and addressing gender imbalances. Most governments in the world have signed up to achieve these goals. The umbrella of sustainable food systems is a multifaceted approach to doing so.

Olivier De Schutter, UN Special Rapporteur on the right to food, states that in his opinion 'Food democracy must start from the bottom-up, at the

The Land Settlement Association *(Marina O'Connell, 2011)*

In the 1930s, in the Great Depression (caused by a banking crisis), there was 70 per cent unemployment in the coal-mining regions of northeast England and South Wales. At that time there was no welfare state to provide housing, unemployment benefits and healthcare. In 1931 an emergency National government was formed to deal with this crisis. One of the solutions was to create the Land Settlement Estates to provide a livelihood for unemployed industrial workers.

The LSA was formed in 1932 and funded by the government and the Carnegie Trust (Figure 11.1). They bought up farms across the country and divided them into groups of 50–80 holdings of 1.6 to four acres each. Each holding was equipped with a piggery, a chicken house, a machinery shed, a glasshouse and cold frames. Approximately 1,000 holdings were created across 21 sites. They were to be run as 'colonies' and miners were 'settled' on the holdings and given basic training to become smallholders. This was viewed as either the first step to emigration to further-flung colonies like New Zealand, or the first step on to the farming ladder. The Estates had central stores and pack-houses and were run as 'compulsory cooperatives'. The tenants had to sell the produce they grew through the

pack-house, and these crops were grown according to a centralised cropping programme for ease of marketing. In this way the LSA hoped to overcome the inefficiencies of the small-sized plots.

By the 1970s the LSA as a whole produced 78 per cent of all the salads in the UK. They were sold not in local markets but in the wholesale markets and were then distributed around the country to the new supermarkets that were springing up. In the early 1980s the tenants were able to buy their holdings and the growers then formed their own cooperatives and have continued trading ever since. But today, 30 years on, the dominance of the supermarkets has all but pushed the last remaining growers out of business. Most now lease out their glasshouses to larger and larger companies that can meet the demand of the supermarkets. Other smallholdings on the estate have been sold on and are now used for horses, as large gardens or, a very few, as market gardens.

When people I interviewed talked about the days when the holdings were in full flow, in the 1960s and 1970s, they used the words 'idyllic', 'utopia', 'paradise' over and over. Hard work yes, but also a time of friendships, community and making a good living. People who were children at that level of villages, regions, cities, and municipalities', and that 'Governments have a major role to play in bringing policies into coherence with the right to food, and ensuring that actions are effectively sequenced, but there is no single recipe.' He goes on to argue that agribusiness is focused on one outcome – yield and profit – and that this needs to change to allow small-scale local agricultural and food systems, including agroecological systems, to flourish.[11] However, the business interests of agribusiness are deeply entrenched, being supported by governments and propped up by the strong lobbying power of giant companies. This is why it is vital that La Via Campesina have a voice in the UN to represent the rights of small-scale food producers.

time told stories of how they 'free-ranged' across the holdings with other children, how their parents were there on site, although very busy. The generation who were the growers in the 1960s and 1970s told me stories of collaboration on the holdings – how they shared knowledge and information freely because they were not competing with each other. One retired grower told me proudly of how they experimented with their crops from a very high skill base and not from an intellectual or college-trained base.

The social clubs and pubs on the estates were the focus of seasonal events such as Bonfire Night, football, darts matches, summer fetes and the annual exhibition of fruit and vegetables with cups and prizes. The growers met up at the central stores on a Friday to collect their supplies and have a chat. The women had a Women's Institute (WI) where they learned or shared the skills they needed for life on a smallholding: jamming, chutney-making, cooking and butchery. Women were not allowed to hold a tenancy in their own right until the 1970s.

In the US in the 1930s the Roosevelt government set up the New Deal, here too creating many cooperative farms, the most famous of which is immortalised by Johnny Cash in his songs. He grew up on a small farm in Dyess, Arkansas, which was one holding on a 'colony' of holdings much like the LSA in the UK.

The LSA experiment lasted 50 years and then was re-privatised. Now these properties change hands in private sales and are worth so much that it does not make economic sense to run them as a food production business.[12]

Figure 11.1: Land Settlement Association. Source: Museum of English Rural Life.

Two of the main barriers to bringing about the change are:

Providing access to land

On a worldwide scale there is no more land that can be brought into food production without catastrophic further loss of biodiversity and environmental services that we rely on. If we examine the problem of the shortage of land for food production, using the permaculture principle of 'small and slow solutions', we find that there are thousands of abandoned small farms across Europe and the US, whose soils are degraded and which are no longer farmed because they are economically unviable. Even on the crowded island of Great Britain there are instances of abandoned market gardens and small farms that are no longer economically viable. In Totnes an abandoned market garden has been paved over to make a car park, just as Joni Mitchell says in her song: 'they paved paradise and put up a parking lot'. Small family mixed farms are often leased to larger and larger companies that can use the economies of scale to produce ever cheaper food for the supermarkets. This trend reduces crop diversity and depends on fewer and fewer skilled people to produce more and more food.

The problem is not only about access to land; it is also about the price of the land, and the need for landowners to give long-term tenancies to farmers, so the farmers can make the financial and labour investment in long-term sustainable food production which will enable them to build economically viable businesses. With adequate routes and access to markets, growing sustainable food can be a fulfilling and rewarding livelihood.

There are examples and models of how land has been brought into food production when it was needed, such as small-scale local food and job creation schemes in the US and the UK in the 1930s in response to the Great Depression. The Land Settlements were created by the government in the UK to rehouse and retrain unemployed miners on smallholdings and played a key role in introducing industrial farming methods. In the US, Roosevelt created farms and smallholdings for the unemployed as part of his New Deal. These examples demonstrate that, if the need is great enough and the political will is there, access to land can be provided simply and quickly. Could this be relevant in the next 30 years with climate change and the need for relocalised and low-carbon food? Currently, land prices are too high and food prices too low for this to make much economic sense, but when food security and access to land are higher on the political agenda, the Land Settlement Association (LSA) might provide a model for establishing a local food economy.

In the global south, land-grabbing of communally owned land and small-scale farms is happening apace for the purposes of agribusiness and biofuel production. These lands and their ownership model must be protected by individual states.

Providing training for sustainable farming systems

Less than one per cent of the population in the UK works in farming or growing and the average age of a UK farmer is in the late fifties, suggesting that a new generation will soon need to take up the task of food production in this country. If we continue with the thought experiment of 50–100 per cent of farms in the world transitioning to sustainable farming methods, how would current or future farmers learn about these methods? Very few young people are going into any form of training for farming, growing or food production. The occupation of being a sustainable food producer or trader needs to be perceived as a profession, possibly of the same status as being an engineer. It is interesting that in Cuba today a farmer has the same status as a doctor. The profession needs to be well paid and young people need access to land and farms.

Training in sustainable food systems is needed at the further education level for skilled workers, at degree level for the managers and creators of these new systems and at postgraduate level for researchers. Sustainable farming and growing and the enterprises that support them, such as farmers' markets and CSA systems, have high uptake by women. The training needs to be made available and accessible to everyone. It needs to be experiential as well as theoretical, and to be state funded.

Farmers can be trained in transitioning to sustainable farming practice – be it agroecology, regenerative, organic, biodynamic, agroforestry or permaculture design – via CPD. The agroecology movement uses farmer-to-farmer training seminars to do this.

Agroecology, agroforestry and regenerative agriculture bachelor's and master's degrees are available in small pockets around the world. Training specifically in organic farming, permaculture or biodynamic farming is rare in any state-funded sector; most of the training available is in the form of apprenticeships on farms or of private short courses run by the various associations. 'Skill-level' training (levels three and four in the UK) in sustainable food systems that is state funded, and therefore accessible and affordable, is practically non-existent. The Apricot Centre at Huxhams Cross Farm launched level three and four training in regenerative food systems in January 2022, and this is state funded.

Extension training provided to farmers by paid government advisors is another method to support farmers to take up sustainable farming methods. This is how the UK government rolled out industrial farming after World War II, and it can be done again to support the transition to sustainable systems.

In conclusion

In summary, it is clear that the methods outlined in this book – biodynamic and organic farming, permaculture, agroforestry, regenerative agriculture and agroecology – offer the knowledge and practices to 'ferment' a sustainable agricultural revolution to produce enough healthy food for a growing population. These systems will increase levels of biodiversity, sequester carbon, reduce carbon inputs to net zero and cope with increasingly erratic weather patterns as the climate changes.

The farms that already use these systems have been slowly created around the world over the last 100 years and exist on the edges of the industrial farming systems that have generated the problems of soil loss and catastrophic biodiversity loss and contributed to climate change and mass health epidemics.

It is possible to make a transition to sustainable farming, equitable trading systems and producing healthy food. The methods to do this are already proven, and the ways to expand and replicate them have been tried and tested. The next step that needs to be taken is to transition the majority of farms to these sustainable methods, to bring about the healing that our earth needs. The changes will be slow, so we need to start them now.

The barriers to the transition to socially just and economically viable and sustainable food systems will be access to land, access to training and access to markets. History offers examples and models of how government policy can support and encourage individual farmers, caterers and consumers to transition to sustainable healthy food. If the will is there, the next generation of sustainable and regenerative farms can be created. We need to begin the task now. To quote Lao Tzu: 'The journey of 1,000 miles starts with the first step.' ∎

1   C. Massy. *Call of the Reed Warbler*, Chelsea Green, Hartford, VT, 2017.

2   FAO, 'Climate Change, Agriculture and Food Security', 2016.

3   https://www.navdanya.org/

4   W. Berry, *What Are People For?*, Random Century, London, 1990.

5   R. Hopkins, *From What Is to What If*, Chelsea Green, Hartford, VT, 2019; http://restauration-bio-durable-mouans-sartoux.fr/

6   https://www.resurgence.org/magazine/article3390-forests-and-freedom.html

7   Massy, *Call of the Reed Warbler*, p. 437.

8   https://thewisdomoftrauma.com/

9   M.J. Rust, 'Climate on the Couch: Unconscious Processes in Relation to Our Environmental Crisis', *Psychotherapy and Politics International* 6(3), 2008, pp. 157–170.

10 See the film *The Power of Community: How Cuba Survived Peak Oil*, 2006.

11 O. De Schutter, 'Democracy and Diversity Can Mend Broken Food Systems: the Final Diagnosis', 2014.

12 Ibid.

Index

Other books from Hawthorn Press

Gardening for Life
The biodynamic way
Maria Thun

Maria Thun is an authority on biodynamics and this book offers a collection of her tried and tested methods, which are the outcome of extensive research into what works and why. Whether you are an experienced gardener or not, whether or not you have used permaculture or grown organic produce before this beautifully illustrated, comprehensive guide is a must for all thoughtful gardeners wishing to work in harmony with nature.

128pp; 210 x 165mm; paperback; 978-1-869890-32-2

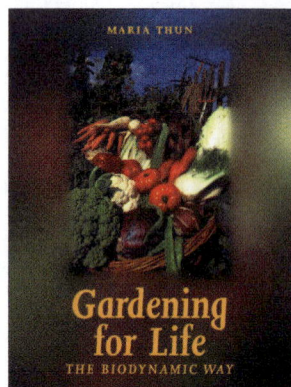

Small Steps to Less Waste
Claudi Williams

Simple ways to make small changes to daily living that will reduce our use of plastics and non-recyclables. This book includes 12 stories of personal enlightenment and simple projects to help you to take back control of your waste and reduce your impact on the environment. It encourages people of all ages to develop the skills to make, create and look after what they have, rather than throw away and buy new.

96pp; 186 x 256mm; paperback; 978-1-912480-29-6

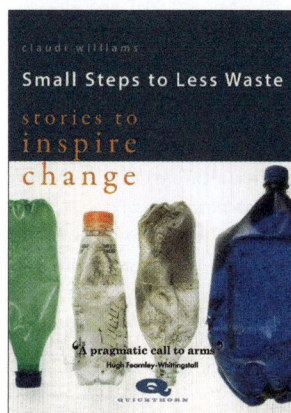

Ordering books

If you have difficulties ordering Hawthorn Press books from a bookshop, you can order direct from our website www.hawthornpress.com, or from our UK distributor:

BookSource: 50 Cambuslang Road, Glasgow, G32 8NB
Tel: (0845) 370 0063, E-mail: orders@booksource.net.

Details of our overseas distributors can be found on our website.

Hawthorn Press
www.hawthornpress.com